Biostatistical Methods

METHODS IN MOLECULAR BIOLOGY™

John M. Walker, Series Editor

METHODS IN MOLECULAR BIOLOGY™

Biostatistical Methods

Edited by

Stephen W. Looney

University of Louisville School of Medicine,

Louisville, Kentucky

Humana Press ✳ Totowa, New Jersey

Production Editor: Diana Mezzina

Cover illustration: Figure 14 from Chapter 4, "Statistical Methods for Proteomics," by Françoise Seillier-Moiseiwitsch, Donald C. Trost, and Julian Moiseiwitsch.

Cover design by Patricia F. Cleary.

For additional copies, pricing for bulk purchases, and/or information about other Humana titles, contact Humana at the above address or at any of the following numbers: Tel.: 973-256-1699; Fax: 973-256-8341; E-mail: humana@humanapr.com; or visit our Website: www.humanapress.com

Printed in the United States of America. 10 9 8 7 6 5 4 3 2 1

Library of Congress Cataloging in Publication Data

Biostatistical Methods / edited by Stephen W. Looney.
    p. cm. — (Methods in molecular biology ; v. 184)
    Includes bibliographical references and index.
    ISBN 0-89603-951-X (alk. paper)
    1. Biometry. 2. Molecular biology. I. Looney, Stephen W. II. Methods in molecular biology (Totowa, N.J.); v. 184.

QH323.5 .B5628 2002
   570'.1'5195--dc21

2001026440

To my mother and the memory of my father

Preface

Biostatistical applications in molecular biology have increased tremendously in recent years. For example, a search of the Current Index to Statistics indicates that there were 62 articles published during 1995–1999 that had "marker" in the title of the article or as a keyword. In contrast, there were 29 such articles in 1990–1994, 17 in 1980–1989, and only 5 in 1970–1979. As the number of publications has increased, so has the sophistication of the statistical methods that have been applied in this area of research.

In *Biostatistical Methods*, we have attempted to provide a representative sample of applications of biostatistics to commonly occurring problems in molecular biology, broadly defined. It has been our intent to provide sufficient background information and detail that readers might carry out similar analyses themselves, given sufficient experience in both biostatistics and the basic sciences. Not every chapter could be written at an introductory level, since, by their nature, many statistical methods presented in this book are at a more advanced level and require knowledge and experience beyond an introductory course in statistics. Similarly, the proper application of many of these statistical methods to problems in molecular biology also requires that the statistical analyst have extensive knowledge about the particular area of scientific inquiry. Nevertheless, we feel that these chapters at least provide a good starting point, both for statisticians who want to begin work on problems in molecular biology, and for molecular biologists who want to increase their working knowledge of biostatistics as it relates to their field.

The chapters in this volume cover a wide variety of topics, both in terms of biostatistics and in terms of molecular biology. The first two chapters are very general in nature: In Chapter 1, Emmanuel Lazaridis and Gregory Bloom provide an historical overview of developments in molecular biology, computational biology, and statistical genetics, and describe how biostatistics has contributed to developments in these areas. In Chapter 2, Gregory Bloom and his colleagues describe a new paradigm linking image quantitation and data analysis that should provide valuable insight to anyone working in image-based biological experimentation.

The remaining chapters in *Biostatistical Methods* are arranged in approximately the order in which the corresponding topic or methods of analysis would be utilized in developing a new marker for exposure to a risk factor or for a disease outcome. The development of such a marker would most likely begin with an examination of the genetic basis for one or more phenotypes. Chapters 3 and 4 deal with two of the most fundamental aspects of research in this area: microarray analysis, which deals with gene expression, and proteomics, which deals with the identification and quantitation of gene products, namely, proteins. Research in either or both of these areas could produce a biomarker candidate that would then be scrutinized for its clinical utility.

Chapters 5 and 6 deal with issues that arise very early in studies attempting to link the results of experimentation in molecular biology with exposure or disease in human populations. In Chapter 5, I discuss many of the issues associated with determining whether a new biomarker will be suitable for studying a particular E-D association. Jane Goldsmith, in Chapter 6, discusses the importance of designing studies with sufficient numbers of subjects in order to attain adequate levels of statistical power.

Chapters 7 and 8 are concerned with genetic effects as they relate to human populations. In Chapter 7, Peter Jones and his colleagues describe statistical models that have proven useful in studying the associations between disease and the inheritance of particular genetic variants. In Chapter 8, Stan Young and his colleagues describe sophisticated statistical methods that can be used to control the overall false-positive rate of the perhaps thousands of statistical tests that might be performed when attempting to link the presence or absence of particular alleles to the occurrence of disease.

Jim Dignam and his colleagues, in Chapter 9, describe the statistical issues that one should consider when evaluating the clinical utility of molecular characteristics of tumors, as they relate to cancer prognosis and treatment efficacy. Finally, in Chapter 10, Greg Rempala and I describe methods that might be used to validate statistical methods that have been developed for analyzing the E-D association in specific situations, such as when the exposure has been characterized poorly.

I would like to express my sincere appreciation to the reviewers of the various chapters in this volume: Rich Evans of Iowa State University, Mario Cleves of the University of Arkansas for Medical Sciences, Stephen George of the Duke University Medical Center, Ralph O'Brien of the Cleveland Clinic Foundation, and Martin Weinrich of the University of Louisville School of Medicine. I am also indebted to John Walker, Series Editor for *Methods in Molecular Biology*, and to Thomas Lanigan, President, Craig Adams, Developmental Editor, Diana Mezzina, Production Editor, and Mary Jo Casey, Manager, Composition Services, Humana Press.

Stephen W. Looney

Contents

Contributors

GREGORY C. BLOOM • *H. Lee Moffitt Cancer Center and Research Institute, Tampa, FL*

JOHN BRYANT • *Department of Biostatistics, University of Pittsburgh and Biostatistical Center, National Surgical Adjuvant Breast and Bowel Project, Pittsburgh, PA*

JAMES DIGNAM • *Department of Biostatistics, University of Chicago, IL, and Biostatistical Center, National Surgical Adjuvant Breast and Bowel Project, Pittsburgh, PA*

ANTHONY. FRYER • *Center For Cell and Molecular Medicine, Keele University, North Staffordshire Hospital NHS Trust, Stoke-on-Trent, Staffordshire, UK*

PETER GIESER • *Net Impact, Inc., Temecula, CA*

L. JANE GOLDSMITH • *Department of Family and Community Medicine, University of Louisville School of Medicine, Louisville, KY*

PETER W. JONES • *Department of Mathematics, Keele University, Keele, Staffordshire, UK*

EMMANUEL N. LAZARIDIS • *H. Lee Moffitt Cancer Center and Research Institute, Tampa, FL*

STEPHEN W. LOONEY • *Department of Family and Community Medicine, University of Louisville School of Medicine, Louisville, KY*

JULIAN MOISEIWITSCH • *Department of Endodontics, School of Dentistry, University of Maryland, Baltimore, MD*

SOONMYUNG PAIK • *Division of Pathology, National Surgical Adjuvant Breast and Bowel Project, Allegheny General Hospital, Pittsburgh, PA*

SUD RAMACHANDRAN • *Department of Clinical Biochemistry, Good Hope Hospital, Sutton Coldfield, West Midlands, UK*

GRZEGORZ A. REMPALA • *Department of Mathematics, University of Louisville, Louisville, KY*

FRANÇOISE SEILLIER-MOISEIWITSCH • *Department of Mathematics and Statistics, University of Maryland, Baltimore Country, Baltimore, MD*

RICHARD C. STRANGE • *Center for Cell and Molecular Medicine, Keele University, North Staffordshire Hospital NHS Trust, Stoke-on-Trent, Staffordshire, UK*

DONALD C. TROST • *Signal Processing, Clinical Technology, Pfizer Global Research and Development, Groton, CT*

PETER H. WESTFALL • *Area of ISQS, Texas Tech University, Lubbock, TX*

S. STANLEY YOUNG • *GlaxoSmithKline, Research Triangle Park, NC*

DMITRI V. ZAYKIN • *GlaxoSmithKline, Research Triangle Park, NC*

1

Statistical Contributions to Molecular Biology

Emmanuel N. Lazaridis and Gregory C. Bloom

1. Introduction

Developments in the field of statistics often parallel or follow technological developments in the sciences to which statistical methods may be fruitfully applied. Because practitioners of the statistical arts often address particular applied problems, methods development is consequently motivated by the search for an answer to an applied question of interest. The field of molecular biology is one area in which this relationship holds true. Even so, growth in application of statistical methods for addressing molecular biology problems has not kept pace with technological developments in the laboratory. Although the story of statistical contributions to the field of molecular biology is still unfolding, a consideration of its history can bring valuable insight into the hurdles—both technical and cultural—still to be overcome in interfacing the two fields. This is especially important given that recent technological advances have created a need for closer interaction among biologists and statisticians. Such considerations also motivated the selection of chapters for this text on statistical methods in molecular biology.

One important question to resolve at the start of such an exploration concerns the name of the field at the intersection of statistics and molecular biology. The term most widely employed by biologists, *bioinformatics*, seems quite inappropriate. The term *informatics* and its derivatives are commonly employed to describe studies of data acquisition and management practices. Evidence of this is the fact that until recently the bioinformatics literature was dominated primarily by computer science applications. Recently its literature has expanded to include areas of what has historically been called *computational biology*. In additional to computer applications, computational biology has

From: *Methods in Molecular Biology, vol. 184: Biostatistical Methods*
Edited by: S. W. Looney © Humana Press Inc., Totowa, NJ

historically focused on the interface of applied mathematics and molecular biology. As described in the following paragraphs, there are substantive reasons to differentiate statistical applications from these, as increased attention is paid to the stochastic nature of data. We considered the adjectival terms *statistical bioinformatics* and *statistical biology*, but discarded these as unsatisfactory owing to their ambiguity. *Statistical genetics* commonly refers to the application of statistical methods to the analysis of allelic data, by which genes directly related to an inherited condition are sought. We discarded this term as being too narrow. Although it includes a biological prefix, the term *biostatistics* has come to refer primarily to applications of statistics to medical research, and more specifically, to the conduct of medical studies. While many studies in molecular biology have eventual medical goals, only a minority have active clinical components. The term *biometry*, which for years has included statistical applications to the biological sciences in its definition, seems much superior. Because an argument can be made that biometry implies an emphasis on the measurement of biological phenomena, we slightly prefer the term *biometric modeling*, which explicitly recognizes the use of inferential models.

Our preference of the term biometry or a derivative such as biometric modeling is vindicated by the clear parallel between the use of images in molecular biology experiments and standard biometric applications such as aerial photographic surveys of wildlife populations. It is further supported by Stephen Stigler, the eminent statistical historian, who points out that by biometry Francis Galton and Karl Pearson meant "the application to biology of the modern methods of statistics" (*see [1]* for a more developed background concerning the field of biometry). Chapters 2 by Bloom et al., and 4 by Sieller-Moiseiwitsch et al., demonstrate especially well the substantive dependence of statistical models on processes of image analysis and quantitation as employed by molecular biologists in laboratory settings. Additional chapters relate biometric applications to biostatistical endeavors in the context of medical studies and clinical trials.

In this chapter we discuss biometric contributions to molecular biology research, and explore factors that have impeded the acceptance of these contributions by the field. We precede this discussion with a historical overview of the growth of molecular biology and of computational methods used therein.

2. Developments in Molecular Biology

Molecular biology encompasses the study of the structure and function of biological macromolecules and the relationship of their functioning to the structure of a cell and its internal components, as well as the biochemical study of the genetic basis for phenotypes at both cellular and systemic levels. Over the last half-century, such research has moved to the forefront of biology and medicine, so much so that molecular biology is sometimes called the "science of life."

Two major classes of questions have been posed by molecular biologists. These concern (1) evolutionary relationships within and across species of organisms and (2) issues of biological functioning in single cells and multicellular systems. Both sets of questions rely on describing and quantifying molecules that serve to characterize types of cells, cellular collections, or organisms, with regard to phenotypes or natural history. Molecules related to interesting phenotypes are called *biomarkers*. This book focuses on measurement and analysis of biomarker quantities because of their importance to our understanding of human disease. The process of identifying important and useful molecular markers is sometimes called *molecular fingerprinting*, *phenotyping*, or *profiling*, particularly when more than one marker is being considered simultaneously.

The idea that molecular fingerprints could be derived for use in characterizing cells and cellular collections has been around for quite some time. In 1958, 2 yr after the sequencing of insulin, Francis Crick recognized that "before long we shall have a subject which might be called 'protein taxonomy'—the study of the amino acid sequences of the proteins of an organism and the comparison of them between species. It can be argued that these sequences are the most delicate expression possible of the phenotype of an organism..." *(2)*. Technological advances over the latter part of the 20th century continuing through the present day have substantially enlarged the availability of molecular data for phenotyping at a number of levels.

Even prior to Crick's remark, peptide fingerprinting techniques, whereby proteins were partially digested into amino acids and peptides and separated by chromatography or electrophoresis, had already been popularized as a means to seek evolutionary similarities across species. The sequencing of insulin, and soon thereafter, of ribonuclease and cytochrome *c*, opened the door to the large-scale characterization of organisms based on protein families. In 1967, the first computer algorithms were developed to seek phylogenetic relationships among a diverse assortment of organisms using a sizable database of protein sequence information. The programs generated binary trees based on a distance metric involving the mutational steps required to move from one protein to another *(3)*. A "residue exchange matrix" was formed from distances between all pairs of measured organisms, and this was employed to represent the overall likelihood of mutations during evolution. In contrast to previous work on molecular data, this approach was very computationally intensive, requiring the use of a digital computer to seek the best binary tree to represent the calculated matrix. Thus was born the field of computational biology.

When Frederick Sanger published the basic chemistry for DNA sequencing in 1975, it became even more apparent that sophisticated database and analytic tools to work with sequence information would be needed. Sanger's approach

Jacob-Monod Central Dogma

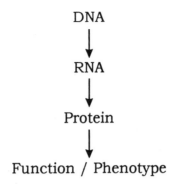

Fig. 1. Central dogma of molecular function and information flow.

is still the primary sequencing technology in use today. In this approach, nested sets of progressively longer DNA fragments are produced. These are then tagged with fluorescent dyes, separated by gel electrophoresis and scanned for identification of basepair sequences *(4)*. Applied Biosystems introduced the first automated sequencing system in 1986. With the addition of robotics to perform the preseparation reaction chemistry, the molecular biology laboratory would become increasingly automated, resulting in corresponding needs for database and analytic tools.

Development of the Southern blot in 1965 and of the Northern blot shortly thereafter marked the passage of another milestone in molecular biology. Biologists use the latter technique to measure the relative amount of RNA message being produced by any specific known gene for which a complementary target has been cloned. By the early 1970s, the basic techniques for studying cellular functioning at the molecular level had been established, so that this functioning could be investigated at each of the levels in the *central dogma* of molecular function and information flow, diagrammed in **Fig. 1**. This states that the direction of cellular information flow is primarily unidirectional, proceeding from DNA to RNA to protein. The genetic program, stored in the DNA, leads to the formation of biological macromolecules—RNAs and proteins—which through their biochemical function result in observable cellular behavior or phenotypes.

Of particular interest to molecular biologists is the manner in which groups of molecules interact to perform a specific function. A set of interacting molecules is said to form a *biological pathway*. **Figure 2** illustrates a particular pathway describing the signaling and functioning of signal transducers and

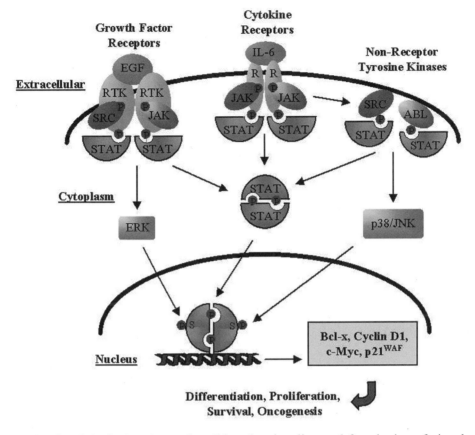

Fig. 2. Biological pathway describing the signaling and functioning of signal transducers and activators of transcription (STATs, *see* **ref. 5**).

activators of transcription (STATs). In brief, STATs are latent cytoplasmic transcription factors that are activated by cytokines, growth factors, and oncogenic tyrosine kinases. They are critical signaling molecules and have been linked to various cellular processes including proliferation, differentiation, and apoptosis. STAT proteins become activated by phosphorylation and dimerization, which allows them to translocate to the nucleus where they bind to specific DNA sequences and in conjunction with other factors control gene transcription. As a result of STAT activation, specific genes are known to be induced that contribute to regulation of cell cycle progression and survival of tumor cells. Experiments and methods that can identify and describe biological pathways such as the one shown in **Fig. 2** are very important to biologists.

The science of *genomics*—the study of the complete set of an organism's genes—was declared when the entire sequence of the bacterium *Haemophilus influenzae* was deduced in 1995. The sequence of baker's yeast, *Saccharomyces cerevisiae*, was completed in 1997. In 1998, a soil-dwelling nematode worm, *Caenorhabditis elegans*, was the first complex animal to have its genome sequenced. The Human Genome Project, with the ambitious goal of mapping the entire human genome in a matter of years, is ongoing.

In spite of this, studies employing one-gene-at-a-time technologies such as the Northern blot form the vast majority of the published investigations up through the present day. In part this is due to the expense and complexity associated with technologies for conducting multigene studies. A more fundamental barrier to the use of comprehensive profiling methods lies in molecular biology training, which emphasizes the full characterization of small networks of interrelated molecules. Recent initiatives by the NIH, such as the NCI Director's Challenge to discover molecular profiles important to cancer biology, seek to change this reticence.

Comprehensive methods for profiling at the levels of RNA and protein matured throughout the 1990s. Chapter 3 presents a history of the development of microarray technology, an extension of Northern blotting whereby an investigator can simultaneously measure the expression of thousands of genes whose complementary sequence has been arrayed on a glass slide or chip. Briefly, the spotted microarray was pioneered at Stanford University in the early 1990s *(6)*. In this implementation, full-length cDNA corresponding to a known gene or an expressed sequence tag (EST) is layered onto a solid surface, usually a treated glass slide, using a robotic arrayer. Total RNA or mRNA is isolated from the sample of interest and labeled cDNA is constructed. The labeled cDNA is hybridized to the cDNA on the surface of the slide and visualized via the incorporated fluorescence tag. Currently, up to 30,000 genes and ESTs can be arrayed on a small glass slide using this technique. A second technology was pioneered by Steven Foder and colleagues in 1991 and has been further developed by Affymetrix *(7)*. The Affymetrix approach uses photolithography and light-activated chemistry to array probes corresponding to different regions of a known mRNA transcript on a solid surface. By combining the signal intensity of the probe sets that query specific transcripts, values for gene expression are obtained. The study of molecular profiles at the level of RNA is sometimes called *transcriptomics*, to parallel the corresponding term at the level of the genome. Chapter 3 by Gieser et al. discusses methods for the design and analysis of microarray studies.

Similarly, *proteomics* is the science that deals with gene end-products, namely, proteins, concerning itself with the set of proteins (the *proteome*) produced by a particular cell, a cellular collection, or an organism. Important information can be derived from experiments seeking to establish whether specific proteins are made in higher or lower concentrations in response to disease, drug treatment, or exposure to toxicants. The most commonly used approach for studying a proteome is two-dimensional (2-D) gel electrophoresis, which combines a first dimension separation of proteins by isoelectric focusing with a second dimension separation by sodium dodecyl sulfate-polyacrylamide gel electrophoresis (SDS-PAGE). The first separation is according to charge (different proteins are focused at their respective isoelectric points) and the second by size (molecular weight). The orthogonal combination of two separations results in a distribution of proteins from a biological sample across the 2-D gel. One or more images of the 2-D gel are collected and analyzed. Chapter 4 by Seillier-Moiseiwitsch et al. covers techniques for finding proteins in 2-D gel images.

To illustrate the potential power of comprehensive measures of gene functioning, consider measurements of mRNA in cancerous and noncancerous tumors or cell lines. Such measurements can elucidate which genes are relatively more or less active in one expression profile as compared to the other, giving an investigator the means to suggest which genes or groups of genes may be important in carcinogenesis. This approach can also be used to understand the effect of drug treatment on a particular tumor, whereby a researcher can investigate which genes change in expression and to what degree as a result of the drug (*see* Chapters 7 by Jones et al. and 9 by Dignam et al.).

3. Developments in Computational Biology

The rapid growth in availability of sequence data at the level of DNA particularly motivated the growth of computational biology. In this context, two basic sets of computational problems were addressed in the late 20th century, both of which were needed to derive possible biological functioning of molecular componentry using computers and mathematics. Each set of problems had implications at every level of molecular phenotyping: genomic, transcriptomic, and proteomic.

The first set of problems depended on the hypothesis that one could predict the functioning of a biological macromolecule if one could only predict its molecular structure. Work in the 1950s and early 1960s had demonstrated that the requisite three-dimensional (3-D) modeling of macromolecules was not feasible using physical, brass-wire models. Cyrus Levinthal demonstrated in 1965 that virtual, computer models of 3-D structures were substantially more amenable to exploration. For two reasons, the elegant theory that the information for the 3-D folding and structure of a protein is uniquely contained in its sequence of amino acids has proven unwieldy in practice. The first impediment has been that various computational complexities arise from its mathematics, whereby solutions require a long series of high-dimensional minimizations often surpassing available computer resources. Although the impact of this impediment could be reduced by providing sufficient correlative information derived from protein crystallographic studies, the cost and difficulty of these studies has been a second major hurdle. To illustrate the early state of this science, consider that the Biological Macromolecule Crystallization Database of the Protein Data Bank (a resource established in 1971 to collect, standardize, and distribute atomic coordinates and other data from crystallographic studies) contained only 12,000 verified protein entries at the time of this writing.

In contrast, perhaps 10 million basepairs are sequenced every day. The wide availability of primary sequence information suggested that tools for rapidly and confidently identifying homologous sequences among those contained in the increasingly large sequence databases could lead to substantive information concerning functioning, without relying on molecular structure. Various search and scoring procedures were developed and applied. For example, in the protein arena, the residue exchange matrix approach referred to previously was modified for use as a scoring matrix in sequence alignment procedures. In this case the matrix is used to determine the likelihood that two residues occur at equivalenced positions in a sequence alignment. Another example relates to the BLAST programs for fast database sequence searching. The name stands for Basic Local Alignment Search Tool. BLAST-style programs use a heuristic search algorithm, seeking to quickly search databases while making a small sacrifice in sensitivity for distantly related sequences *(8,9)*. Databases are compressed into a special format, and the program compares a query sequence to each sequence in the database in the following manner. Sequences are first abstracted by listing exact and similar words within them. BLAST uses these abstracted words to find regions of similarity between the query and each database sequence, after which the words are extended to obtain high-scoring sequence pairs (HSPs). This approach was extended to include gaps in the alignments (GAPPED BLAST), and combined with the scoring matrix approach to increase the sensitivity of hits (PSI-BLAST).

The two approaches to predicting biological function of a molecule—based on its molecular structure or on its primary sequence—could also be combined. For example, it was noted that evolutionary mutations are not equally likely to occur at different positions in a protein, and that a single scoring matrix for all positions in the sequences to be aligned may be inadequate. Overington et al. *(10)* extended Dayhof's idea by using multiple matrices to reflect different mutation probabilities in different regions of a sequence. Similarly, Bowie et al. *(11)* created an exchange matrix for each position in the sequence. Inhomogeneous scoring approaches such as these require reference to a three-dimensional protein structure for at least one of the family members, in order to estimate the parameters of the exchange matrix. The discovery process could also be reversed. Sander and Schneider *(12)* described a procedure to estimate protein structure based on sequence profiles derived from multiple sequence alignments.

Toward the end of this chapter we mention several statistical approaches to sequence alignment and search problems as well as to problems of structural modeling; however, these approaches have had relatively little impact on molecular biology practices to date. In the next section we explore one barrier to the integration of statistical thinking into molecular biology practice.

4. Statistical Content of Academic Programs in Computational Biology

The growth of research in the field of computational biology was accompanied by the initiation of corresponding academic programs. As of October, 2000, the International Society for Computational Biology listed 44 university programs in bioinformatics and computational biology, 29 of these at 23 universities in the United States and Canada. Of these, 13 offer degree programs at the undergraduate or graduate level with required curricula in bioinformatics or computational biology.

To understand the curricular content of these training programs we created an *ad hoc* scoring system, which we used to rate the focus of the programs on a standardized, multidimensional scale. Curricula were obtained from eight graduate programs in the United States. Explicitly required upper-level prerequisite courses were included in this analysis. Each course from each curriculum was categorized according to whether its primary focus was most closely aligned with statistics, nonstochastic mathematics, physics/imaging sciences, biology, medicine, computer science, medical informatics, or bioinformatics. Bioinformatics courses were those that evidenced a multifaceted syllabus including biological databases, sequence searching and alignment technologies, and general techniques for genomics, transcriptomics, or proteomics. In the "other" category we included general seminars and courses on such topics as law and ethics. We based our assignments primarily on course syllabus, course title, the listing department or its code, and lastly on the training and interests of the primary faculty instructor when available over the web. Elective courses were valued at half the weight assigned to required courses in a curriculum, to reflect their relatively lower impact on training. Required upper-level prerequisite courses were given the same weight as required courses.

Although our scoring system is admittedly arbitrary, the results of our analysis are interesting in that they demonstrate that the statistical training offered to students in these programs is secondary to the trinity of required and recommended biology, mathematics, and computer science coursework. On average, statistics coursework accounted for about 10% of the curriculum of these training programs (range: 5–25%). The bulk of required statistics courses were either introductory or focused exclusively on probability theory. **Table 1** summarizes our analysis of the average curricular focus for these graduate programs.

Table 1
Average Percentage of Computational Biology
Curricula Devoted to Various Subject Areas

Primary course focus	Percentage of curriculum
Biology	37.4%
Mathematics	14.0%
Computer Science	13.6%
Biostatistics/statistics	10.5%
Bioinformatics	9.8%
Physics/imaging sciences	3.4%
Medical	2.5%
Medical informatics	2.1%
Other	6.7%

The fact that, to date, the field of statistics has had relatively little impact on the practice of molecular biology is not unrelated to its relative underemphasis in computational biology coursework. Not only do regular biology programs provide even less quantitative training, but the most quantitative, modeling-oriented subset of biology is infused with a constructionist philosophy that contrasts sharply with the reductionist training that is the hallmark of statistics programs. *Constructionists* prefer the creation of reasonable models from consideration of the underlying science to immediate consideration of data, while *reductionists* prefer to generate a model from observed data, perhaps illumined by a small subset of scientific considerations, than to deal with the full-scale intricacies of the underlying science. The constructionist approach largely informs both mathematics and computer science training. Although it is true that computer scientists also perform "data mining" tasks, typical computer science approaches to data-driven analysis tend toward "black box" methods. Neural networks are prime examples of black box methods, as they typically result in little or no acquisition of generalizable, structural knowledge. It has been noted that there are benefits to be derived from the integration of both philosophies into applied work *(13)*. Without the stochastic component, however, quantitation of uncertainty in inferential results is not possible. We believe it is essential that more statisticians be encouraged to support molecular biology studies, another key reason for providing this book as an introductory reference. As we describe in the next section, such statisticians will be rewarded by rediscovering the roots of their field.

5. A Brief History of Statistical Genetics

Having suggested that computer science and constructionist modeling approaches inform computational biology in practice, it is worthwhile to note that the history of statistics in evolutionary biology is a long one. The term *statistical genetics* has come to refer to the application of statistical methods to the analysis of allelic data, by which genes directly related to an inherited condition are sought.

Starting in the early 1870s, Francis Galton became widely known for his championing of eugenics, the science of increasing human happiness through the improvement of inherited characteristics. The creation of a science of eugenics required of Galton that he attempt to solve a number of complex problems, including that of how hereditary traits were transmitted by reproductive processes. Galton's energies gradually focused on statistical reasoning about hereditary processes. This work informed the so-called "Biometrical" school of thought, which believed that continuously varying traits exhibited *bleeding* inheritance. With the rediscovery of Mendel's work on the genetics of dichotomous traits in the early 20th century, fierce debates between Mendelians and biometricians focused on whether discrete and continuous traits shared the same hereditary and evolutionary properties. It is well known that this clash was influenced more by personalities than by facts, but it did serve to motivate further development in the statistical thinking needed to address genomics questions.

By the 1920s, the basic ideas of statistical genetics were developed by Fisher and Wright. These ideas formed a synthesis of the statistics, Mendelian principles and evolutionary biology needed to extend genomic modeling to continuously varying characteristics, which may or may not also exhibit polygenic (multilocus) etiology. These ideas were almost immediately embraced by plant and animal breeders. Extension into human models was developed theoretically, but its rewards would await later developments in computer science and molecular biology. The study of statistical genetics laid the foundation for many advances in theoretical and applied statistics, such as regression and correlation analyses, analysis of variance (ANOVA), and likelihood inference.

An excellent survey article by Elston and Thompson *(14)* breaks the study of statistical genetics into four major areas—population genetic models, familial correlations, segregation analysis, and gene mapping. Chapters 6–9 discuss some of these approaches. We note the availability of a software package called SAGE (Statistical Analysis for Genetic Epidemiology), a collection of more than 20 programs for use in genetic analysis of family and pedigree data. SAGE is available through the Human Genetic Analysis Resource of the National Center for Research Resources.

6. Biometric Modeling: Interfacing Molecular Biology and Statistics

There are three general areas in which molecular biology and statistics are interfacing at the present time. The first is in the context of clinical trials using molecular biomarkers. Most standard methodologies for the design and execution of clinical trials apply in this context. Chapter 9 by Dignam et al. in particular illustrates the use of a biomarker in a clinical trials setting. This setting is somewhat complicated by the fact that many such clinical trials involve multiple biomarker studies, resulting in little conclusive power for individual tests. An additional complication arises from the fact that many molecular biomarkers are quantitative summaries derived from images. Inter- and intraobserver variability associated with biomarkers is frequently studied prior to the initiation of a clinical trial, with methods such as those described in Chapter 5 by Looney; however, variability and bias resulting from image analysis is frequently ignored once a biomarker has entered the clinical trials situation. Chapter 2 by Bloom, et al. suggests a means whereby such information could be incorporated into these studies, increasing their generalizability.

The second area in which molecular biology and statistics are interfacing is in the context of the standard questions of computational biology, described previously, related to the functioning of biological macromolecules. Various statistical models have been proposed to perform sequence alignment *(15–20)*. The work of Charles Lawrence and his colleagues is particularly interesting, in that they have employed Bayesian methods to address these problems. For example, the Bayesian solution to a product multinomial model has been proposed to perform multiple alignment, detecting subtle sequence motifs shared in common by a given set of amino acid or nucleotide sequences. One such model employs a Bernoulli motif sampler which assumes that each sequence could contain zero or more motif elements of each of a set of motif types. Starting with an alignment of motifs, the site sampler proceeds to follow two Gibbs sampling steps. First is a predictive update step that chooses one of the N sequences in order from first to last. The motif element for each motif type in the chosen sequence is added to the background and counts of discovered motifs are updated. Second is a sampling step, in which the probability associated with each possible motif starting position is estimated according to a model. Weighted sampling of a single motif element is then conducted for each motif type. This two-step process is repeated until a local maximum alignment has been obtained. Bayesian models have also been employed in the context of protein folding *(21,22)* and RNA structure *(23)*. Although formal statistical modeling in problem areas in the traditional domain of computational biology has had little impact on molecular biology practice thus far, the fact that statistical model-based search methods are found to provide substantial improvement over current non-model-based methods *(20)* bodes well for the future.

A third area of interaction between statistics and molecular biology is slowly emerging because of recent advances in comprehensive, high-throughput laboratory methods for studies of gene expression at the levels of RNAs and proteins. Chapter 3 discusses methodological issues and approaches related to studies employing microarray technology, and Chapter 4 discusses approaches for 2-D protein gel analysis.

7. Conclusions

The infusion of statistical methods into the field of molecular biology promises to substantially enhance current scientific practices. Improved tools resulting in superior inference may be required to ensure important scientific breakthroughs. In this chapter we summarized statistical work in relation to the fields of molecular and computational biology, and explored some of the barriers still to be overcome. This book seeks to assist in the fusion of statistics and molecular biology practice by focusing on methods related to biomarker studies and molecular fingerprinting. We hope that it will prove useful to statisticians and biologists alike.

References

1. Stigler, S. (2000) The problematic unity of biometrics. *Biometrics* **56,** 653–658.
2. Crick, F. H. C. (1958) On protein synthesis. *Symp. Soc. Exp. Biol.* **12,** 138–163.
3. Dayhoff, M. O. (1969) Computer analysis of protein evolution. *Sci. Am.* **221,** 87–95.
4. Sanger, F., Nicklen, S., and Coulson, A. R. (1977) DNA sequencing with chain-terminating inhibitors. *Proc. Natl. Acad. Sci.* **74,** 5463–5467.
5. Bowman, T., Garcia, R., Turkson, J., and Jove, R. (2000) STAs in Oncogenesis. *Oncogene* **19,** 2474–2488.
6. Schena, M., Shalon, D., Davis, R. W., and Brown, P. O. (1995) Quantitative monitoring of gene expression patterns with a complementary DNA microarray. *Science* **270,** 467–470.
7. Lockhart, D. J., Dong, H., Byrne, M. C., Follettie, M. T., Gallo, M. V., Chee, M. S., et al. (1996) Expression monitoring by hybridization to high-density oligonucleotide arrays. *Nat. Biotechnol.* **14,** 1675–1680.
8. Altschul, S. F., Gish, W., Miller, W., Myers, E. W., and Lipman, D. J. (1990) Basic local alignment search tool. *J. Mol. Biol.* **215,** 403–410.
9. Karlin, S. and Altschul, S.F. (1990) Methods for assessing the statistical significance of molecular sequence features by using general scoring schemes. *Proc. Natl. Acad. Sci. USA* **87,** 2264–2268.
10. Overington, J., Donnelly, D., Johnson, M. S., Sali, A., and Blundell, T. L. (1992) Environment-specific amino acid substitution tables: tertiary templates and prediction of protein folds. *Protein Sci.* **1,** 216–226.
11. Bowie, J. U., Luthy, R., and Eisenberg, D. (1991) A method to identify protein sequences that fold into a known three-dimensional structure. *Science* **253,** 164–170.
12. Sander, C. and Schneider, R. (1991) Database of homology-derived protein structures and the structural meaning of sequence alignment. *Proteins* **9,** 56–68.

13. Lazaridis, E. N. (1999) Constructionism and reductionism: two approaches to problem-solving and their implications for reform of statistics and mathematics curricula. *J. Statist. Educ.* [Online] **7**, *http://www.amstat.org/publications/jse/secure/v7n2/lazaridis.cfm.*

14. Elston, R. C. and Thompson, E. A. (2000) A century of biometrical genetics. *Biometrics* **56,** 659–666.

15. Gribskov, M., McLachlan, A. D., and Eisenberg, D. (1986) Profile analysis: detection of distantly related proteins. *Proc. Natl. Acad. Sci. USA* **84,** 4355–4358.

16. Lawrence, C. E., Altschul, S. F., Bogouski, M. S., Liu, J. S., Neuwald, A. F., and Wooten, J. C. (1993) Detecting subtle sequence signals: a Gibbs sampling strategy for multiple alignment. *Science* **262,** 208–214.

17. Krogh, A., Mian, I. S., and Haussler, D. (1994) A hidden Markov model that finds genes in *E. coli* DNA. *Nucleic Acids Res.* **22,** 4768–4778.

18. Liu, J. S., Neuwald, A. F., and Lawrence, C. E. (1995) Bayesian models for multiple local sequence alignment and Gibbs sampling strategies. *J. Am. Statist. Assoc.* **90,** 1156–1170.

19. Neuwald, A., Liu, J., Lipman, D., and Lawrence, C. (1997) Extracting protein alignment models from the sequence data database. *Nucleic Acids Res.* **25,** 1665–1677.

20. Park, J., Karplus, K., Barrett, C., Hughey, R., Haussler, D., Hubbard, T., and Chothia, C. (1998) Sequence comparisons using multiple sequences detect three times as many remote homologues as pairwise methods. *J. Mol. Biol.* **284,** 1201–1210.

21. Jones, D. T., Taylor, W. R., and Thornton, J. M. (1992) A new approach to protein fold recognition. *Nature* **358,** 86–89.

22. Bryant, S. H. and Lawrence, C. E. (1993) An empirical energy function for threading protein sequence through the folding motif. *Proteins* **16,** 92–112.

23. Ding, Y. and Lawrence, C. E. (1999) A Bayesian statistical algorithm for RNA secondary structure prediction. *Comput. Chem.* **23,** 387–400.

2

Linking Image Quantitation and Data Analysis

Gregory C. Bloom, Peter Gieser, and Emmanuel N. Lazaridis

1. A Shifting Paradigm

Until recently, image-based experimentation in molecular biology has been primarily concerned with qualitative results produced as a result of such experiments as Northern blots, immunoblotting, and gel electrophoresis. These experiments result in a relatively small number of bands on an autorad or other imaging medium. These bands or spots would be visually inspected to determine their "presence" or "absence," or visually compared with other spots on the medium to determine their relative intensities. Sometimes, comparisons would be enhanced using quantities derived from densitometry analysis. Such comparisons were often performed to provide a numerical summary of a clearly visible difference. This summary may have been required for publication of the experimental results. This approach seemed to serve the investigator well because there existed no real need for accurate image quantitation or data analysis and a simple qualitative result would suffice.

However, many recent advances in molecular biology, coupled with the increasing knowledge of the human genome, have made possible the ability to simultaneously test the expression level of several thousand individual genes, as in the case of microarray analysis (*see* Chapter 3 by Gieser, et al.), or hundreds of expressed proteins, as with two-dimensional (2-D) gel electrophoresis (*see* Chapter 4 by Seillier-Moiseiwitsch, et al.). While this ability is essential to further molecular biology research and is a giant leap forward from more traditional approaches, it has raised several questions about the use of the "old" paradigm of image quantitation and data analysis and whether that paradigm can be successfully applied to these new image types. Several characteristics of modern molecular biology experiments—including the need to investigate and understand subtle changes in molecular quantities and the

From: *Methods in Molecular Biology, vol. 184: Biostatistical Methods*
Edited by: S. W. Looney © Humana Press Inc., Totowa, NJ

increasing sensitivity of quantitation to the imaging process—suggest that the old paradigm must be modified. In this chapter, we suggest a new approach that allows investigators to better handle the needs of image-based experimentation.

To demonstrate why a new paradigm linking image quantitation and data analysis is needed, and to better understand the scope of the problems faced when analyzing a laboratory image, we briefly describe some of the new technologies and the image types they produce.

Microarray analysis (*see* Chapter 3 Gieser, et al.) is a procedure that allows an investigator to simultaneously visualize the expression levels of thousands of genes whose complementary sequence or a portion thereof has been arrayed on a class slide or chip. The measurement of mRNA levels in, for instance, a normal tissue or cell line to its paired experimental sample can elucidate which genes and, indirectly, which proteins are present or absent, and their relative expression levels in one condition as compared to another. This gives the investigator a starting point to determine which genes or groups of genes are important in a particular experimental context. Regardless of the type of question(s) being asked, this experiment invariably results in a large image or set of images with thousands of features, each of which needs to be geometrically defined into a region of interest (ROI) and subsequently quantitated. The microscopic scale on which this kind of experiment is performed plays an important role in determining the sensitivity of analytic results to the imaging process. Ratios of quantities across images are frequently needed to compare the relative expression across conditions.

The second type of modern biological image-based experimentation is termed *proteomics*. This is the science that deals with gene products, namely, proteins, and concerns itself with the collection of proteins (the "proteome") produced by a particular cell or organism. Important information can be derived from experiments seeking to establish whether specific proteins are made in higher or lower concentrations in response to disease, drug treatment, or exposure to toxicants. The most commonly used approach to protein identification and quantitation is 2-D gel electrophoresis, which combines a first dimension separation by isoelectric focusing (IEF) with a second dimension separation by sodium dodecyl sulfate-polyacrylamide gel electrophoresis (SDS-PAGE). Whereas microarray experiments result in images with features whose geometry is determined by the physical assembly of samples on a substrate, proteomics 2-D gel images consist of many spots whose location and shape cannot be prespecified easily. As with the microarray, the 2-D gel image consists of several hundred to several thousand features of varying intensity that need to be characterized. Each feature may or may not be important in the context of a given experiment.

In both microarray and proteomics 2-D gel contexts, effects due to image background, signal-to-noise ratio, feature imaging response and saturation, and experimental design and execution must be accounted for and factored into the overall image quantitation procedure. Any and all of these factors can have far-reaching effects on the subsequent data analysis. Under the old paradigm, determination of the effects of variation in these factors on subsequent data analysis is impossible, if for no other reason than that image quantitation would typically proceed under a single set of conditions, in a step that would never be revisited. If subtle differences in the performance of image quantitation may substantially affect the subsequent data analysis, then the old paradigm simply no longer serves, as it allows only a single best "guess" at what imaging parameters are optimal and allows for no testing to see if the guess was correct.

A key point of the discussion of our new paradigm for treating images from biological experimentation is that image quantitation can have a potentially large effect on the data that are being obtained and these effects would feed through the subsequent data analysis. In any imaging experiment there exists an infinitely large number of ways in which an image can be quantitated, all of which may be "correct" in that they all lie in some reasonable envelope of imaging procedures. Among these methods, and even across subtle variations of a single method, substantial variability in quantitation may result. This is particularly important when one considers searching for subtle trends or effects in a data set. For the newer types of image-based biological experimentation, such subtle differences in how image quantitation is performed can completely alter the data analysis outcome. A method is needed for linking the imaging and data analysis processes so that the one can feed into the other, enabling the investigator to understand the effect of choices made in image quantitation on the resulting data analysis. The reverse situation is also important, as the results of data analysis can drive the choice of procedures for image quantitation. For example, an analysis of data derived from a particular procedure for image quantitation using a specific background cutoff value in a given image may demonstrate that the imaging procedure eliminated too many features of the image from consideration, necessitating that the image quantitation process be revisited. The idea of using the results of one of the two steps in this process to drive the other process is central to the new paradigm.

Such an approach is important for the analysis of the newer types of biological images produced today because of their sheer complexity and the large number of features contained within each image. In hindsight it seems that the traditional segregation of image analysis from the data analytic process may have been sub-optimal for analysis of more traditional types of biological experiments as well.

In this chapter, we introduce a new paradigm with an accompanying schema for the treatment of experimentation involving images and their subsequent data analysis, and point out the benefits of this new approach. The new approach encourages cooperation between image quantitation and data analysis. Ideally, this implies that the two processes should be performed by a single software application. While not necessary, integration of imaging and statistical software tools can make application of the new paradigm easier, as we will describe and illustrate in detail later in the chapter.

2. Conceptualization of the New Paradigm

The first action with any imaging experiment is to produce the medium with the features or items to be imaged. The medium can be a microarray chip or slide, a proteomics 2-D gel, or any number of other experimental media. The second step is image acquisition. This can be as simple as scanning a piece of exposed film or as complex as scanning a 2-D SDS-PAGE gel in a proteomics experiment. To illustrate the use of our novel paradigm for the image quantitation and data analysis step of an imaging experiment, a workflow diagram is shown in **Fig. 1**.

The first step in this process is image quantitation. Image quantitation consists of translating the underlying pixel information in the image into useful data through the use of imaging methods. The set of imaging methods and their associated parameters constitute an *imaging envelope*. Methods in the imaging envelope may differ in how they treat background signal information, identify signal in the presence of noise, characterize feature geometry, and identify features with labels. The parameters that are required by a particular method to perform quantitation can include numerical summaries of background signal, expected signal-to-noise ratios, or signal thresholds for the image. The methods and parameter values defining the imaging envelope are what determine the values of the resulting *data sets* (shown below the imaging envelope in **Fig. 1**). It is important to note here that even subtle changes in the imaging envelope can lead to large changes in the acquired data set(s). These changes will, in turn, alter the *inferences* obtained by application of the *statistical algorithm*. It is therefore very important to incorporate a reality check after data analysis and a subsequent feedback mechanism for improving specification of the imaging envelope. Modifying the envelope in turn will necessarily alter the inferences. Note that in some situations, particularly when formal analytic protocols must be consistent over multiple analyses, feedback may be undesirable beyond an exploratory stage.

After the data sets are obtained, a single statistical algorithm is applied to each individually. The type of statistical algorithm used is not critical to the paradigm and may be anything from a *t*-test to linear regression. In one possible path of workflow, the inferences are grouped into a *inference set*, representing the individual values obtained from application of the statistical algorithm. At

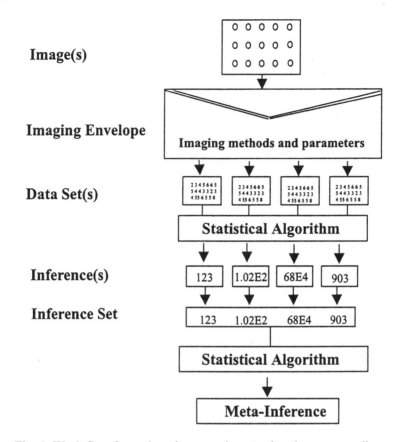

Fig. 1. Work flow for an imaging experiment using the new paradigm.

this point a meta-analysis of the inference set, using analysis of variance (ANOVA), for example, is performed to arrive at a summary description or *meta-inference*. This summary result incorporates not only the final outcome of the data analysis, but also a measure of the variability or potential error introduced by the imaging process.

The other possible path through the work flow diagram summarizes the data sets obtained as a result of image quantitation into a single meta-data set before application of a statistical algorithm. This treatment leads to a single inference at this point in the flow and no further analysis is necessary. This approach has an advantage in that it is more amenable to specification of distributions for parameters characterizing the imaging envelope. For example, when one is interested in integrating out the effect of a particular parameter from a specific imaging algorithm, one can place a prior distribution on that parameter and calculate an inferential posterior distribution using a Bayesian approach. In addition, the loss of information resulting from the application of the statistical algorithm occurs at only one point in the flow, making it easier to evaluate goodness-of-fit. Disadvantages of this approach include the fact that the image quantitation procedures must be "compatible" across the imaging envelope so they can be combined in the context of conducting a single statistical operation. In the alternative approach, only the intermediate inferences, and not their underlying data sets, need be combined for the subsequent statistical analysis leading to the meta-inference. Therefore, different statistical algorithms may be applied to each individual data set as long as the inferences can be meta-analyzed, making this approach more flexible.

When the process illustrated in **Fig. 1** is integrated in a single software platform, models of the experiment that account for use of different imaging parameters and quantitation procedures can be more readily explored, reducing the potential for imaging-related biases in the analytic results. The sensitivity of any given analysis to changes in quantitation procedure can also be rapidly investigated, thereby increasing the quality of information derived even from simple statistical models. The next section describes a novel application that allows this conceptual solution to be practiced in a real-world environment.

3. Application of the Paradigm: The Midas Key Project

While it is easy to conceptually cycle through several rounds of quantitation and data analysis using the approach described in **Subheading 2.**, it is much more difficult to perform this task in a real-world environment. This is especially true if the processes of image quantitation and data analysis are physically separated. In fact, this is the situation that currently exists. Many systems are available for image analysis, including home-grown and commercial, general and special-purpose packages such as Optimas (Media Cybernetics, Inc.; general purpose imaging), SpotFinder (TIGR; microarray slide imaging), and CAROL (Free University of Berlin; proteomics 2-D gel imaging). Indeed, many vendors of biological equipment produce and distribute their own software, which they bundle with their equipment. While some of the available packages may provide sophisticated image-analysis tools, little sophistication is available in the included mathematical and statistical methods for analysis of the resulting data. Conversely, popular analysis packages such as SAS, SPSS, and S-Plus, while providing sophisticated models for data analysis, lack any facility for image quantitation. Thus, the typical scientific segregation of the analytic role from the process by which image-related data are obtained is also reflected in available software. While such software may suffice to conduct the kinds of traditional biological experimentation that relied primarily on qualitative examination of images, it was recognized that use of such software in the context of the new biological experimentation would be suboptimal.

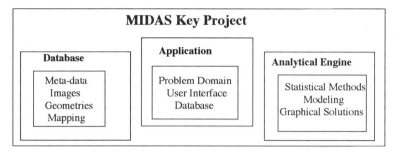

Fig. 2. Diagram of the three components constituting the MIDAS Key Project and their elements.

The paradigm of marrying imaging and mathematical modeling and statistical tools to analyze the results of modern biological experimentation could be implemented using the disparate software applications described previously. This approach has several limitations, however, foremost of which is the ability to quickly incorporate the results derived from either of the two analytic domains into the other. One would need to go back and forth between the imaging and data analysis exercises hundreds or thousands of times. A platform that would allow imaging and data analyses to proceed in tandem would substantially enhance the analytic exercise. Thus, we have been developing an application that incorporates all aspects of image and data analysis along with data storage into a single unit. We describe the design and merits of this application in the paragraphs that follow.

The goal of the MIDAS Key Project is to build an integrated imaging and modeling analytic environment over a sophisticated database backbone. By borrowing and uniting technologies from multiple fields, we seek to empower researchers in basic and clinical imaging studies with a sophisticated analytic toolbox.

Figure 2 illustrates the three major components that constitute the MIDAS Key Project. A description of each of the elements contained in each of the components is also given. The top-level box is the Java *application*. This is the central component of the key project and ties the other components together. The Problem Domain of the application contains the objects that define the underlying data structures used for the project. The Java application also controls interactions with the database; this is done in the Database area. The third area is the User Interface. This package is responsible for all aspects of interaction with the application, including the menu-driven frame-based interface and image display. The use of Java allows us to maintain cross-platform independence; to integrate tools existing in multiple, otherwise unrelated, applications; and to easily deploy a client-server multithreaded model system.

We are currently using Java 2 as the basis for our code, supplemented by the Java Advanced Imaging (JAI) Application Programming Interface (API). The JAI API is the extensible, network-aware programming interface for creating advanced image-processing applications and applets in the Java programming language. It offers a rich set of image processing features such as tiling, deferred execution, and multiprocessor scalability. Fully compatible with the Java 2D API, developers can easily extend the image-processing capabilities and performance of standard Java 2D applications.

The current Java Development Kit (JDK) fully incorporates Swing components (which are used for windowing functions) and the 2D API, both of which are employed throughout our code. The Java Database Connectivity (JDBC) API allows developers to take advantage of the Java platform's capabilities for industrial-strength, cross-platform applications that require access to enterprise data.

The *database* component of the MIDAS Key Project contains the table spaces that hold all long-term storage needed in the application. The tables contained here include those for storing project, experiment, and image metadata; tables to store the images and their associated geometries; and mapping tables to tie the data together. For our work we chose to employ Oracle for all data storage and management. Oracle provides many unique technical features that we leverage in the Key Project including Java integration, extensibility and scalability, and support for multimedia data types that allow for efficient integration of imaging and meta-data information.

The most important characteristic of an *analytical engine* in the Key Project is its amenability to integration with other software, including novel statistical methods. A second characteristic is the ease with which it interacts with Java applications. We chose to employ the S-Plus statistical processing system for our work in spite of the fact that it is not fully Java aware. A fully Java-aware analytic engine would allow dynamic statistical methods to be incorporated into our Java interfaces, allowing application of real-time graphical data exploration methods and interactive statistical diagnostics. In addition, we can conveniently employ the S-Plus system on desktop computers separately from our Java interfaces, assisting in rapid methods development and evaluation.

Figure 3 shows a typical application of the Key Project system, focusing on the Oracle backbone, which is used for object persistence. First, a series of image-dependent or imageless layers, upon which analysis will be performed, are loaded into the system (step 1). Memory is carefully managed at this step and throughout the process, as it is impossible to expect either client or server to simultaneously manage, say, 40 microarray images, each of which is upwards of 40 Mb long. A rendered composite image, if available, is displayed on the client according to user-adjustable preferences. We allow for imageless layers so that we might work in our analytic environment with data obtained

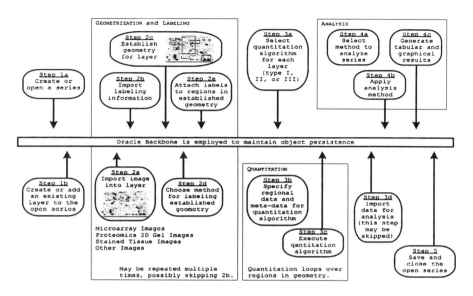

Fig. 3. Schematic of the MIDAS Key Project system showing Oracle backbone.

through sources whereby the associated images are not available. When images are available, we proceed to establish one or more geometries for each layer (step 2). By a *geometry* we mean a set of closed, possibly overlapping regions-of-interest (or shapes), each of which is not exclusively contained in any other. Geometry may be established by hand through a sketchpad interface or by application of a geometrization algorithm. The use of geometrization algorithms allows us to model in a single system images with formats that are largely fixed by the investigator, such as, for example, results from microarray studies, images with semifixed geometries such as from proteomics studies, and images with free-form geometries such as from cell or tissue microscopy. Labels are then attached by reference to one or more labeling algorithms (end of step 2). These may be relatively simple—typically, microarray labels are established by considering the spot centers—or fairly complex—protein labels on 2-D gels are established by considering the overall geometry and relative positions of shapes in that geometry. Geometries are calculated and labels established. Next, quantitation is carried out (step 3) by referencing one or more quantitation algorithms, which execute looping over shapes in the geometry. Quantitation may result in all kinds of information, including: (1) primary signal information, such as average or median intensity of the pixels in regions of interest; (2) signal variability information, such as pixel variance, kurtosis, or direction of one or more principal components; (3) signal location information, such as coordinates of the intensity mode within a region of interest; and (4) cross-image signal comparison information, such as pixel correlation between two images (used for quality control). The design of our system allows for substantial extensibility in the application of geometrization, labeling, and quantitation algorithms. Depending on the algorithm, quantitation may be performed by server-side Java or C++ code or by the S-Plus Server system. Note that geometrization algorithms may also be employed within the quantitation step, without requiring persistent storage of the resulting geometry, as might be needed when one wishes to compare quantitative performance of two spotfinding algorithms within regions of interest in a specified geometry. External data, for which no images are available, are also retrieved at this time. Analysis of the quantitation results occurs in step 4. We employ standard methods such as simple regressions, ANOVA, and principal components analyses by referring to the methods built into the S-Plus analytic engine. Novel mathematical models are included by incorporating C++ or Fortran compiled code into the S-Plus engine or by direct reference to external code on the server. Graphical, tabular, or data-formatted results can be exported for reports or stored on the Oracle backbone for later use (step 5).

Initial exploration of multiple image-based experiments suggests that the variability associated with application of reasonable but differing imaging procedures to the same images is nontrivial. The total effect this variability will have on various statistical models is unknown at present. Without reference to our new paradigm for imaging and data analysis, it would remain largely unknowable.

4. Midas Center at USF—An Interdisciplinary Implementation of the New Paradigm

The new paradigm has changed the way researchers at our institution interact to analyze imaging-based experiments. The University of South Florida (USF) Center for Mathematical-Modeling of Image Data Across the Sciences (MIDAS) brings together faculty and student investigators from disparate fields to develop sophisticated mathematical and statistical models of data derived from images. Under the umbrella of MIDAS, we seek to address pressing analytic needs related to molecular biology experiments in many areas, including microarray, microscopy, proteomics, and flow cytometry. In each kind of experiment, an image or a set of images is typically derived by a primary investigator—say, a biologist or pathologist—in an experimental context. To get from the images to informative research conclusions, the steps of quantitation, analysis, and interpretation must be traversed. In today's research environment, the primary investigator usually directs quantitation of the images, sometimes in conjunction with an imaging scientist. The resulting data may then be given to a statistician or other numerical analyst. The basic tenet of the MIDAS Center is that segregation of the analytic role in this context is suboptimal. At the present time, the MIDAS Center is integrating researchers from multiple schools and programs around USF. Investigators from programs in Biology, Bioengineering, Computer Science, Mathematics, Statistics, Medicine, Medical Imaging, Oncology, Biochemistry, Pathology, and Public Health are collaborating to address important analytic problems.

5. Example

Synchronous implementation of the new paradigm in both software and the collaborative environment allows for easy conduct of joint imaging and analysis experiments. In this section we first present a hypothetical experiment employing the new paradigm, and then illustrate application of the paradigm to an image using the MIDAS Key Project.

A hypothetical experiment using the new paradigm might be the following. Suppose we have conducted an experiment using 40 microarray slides that were assembled on two different days. We are concerned that our data analysis might be sensitive to problems we suspect with the microarrayer pins, and we have developed three combined sets of geometrization, labeling, and quantitation algorithms that we can apply to these data, each of which has some benefits and some drawbacks in terms of ability to adjust the resulting data for experimental difficulties. Each algorithm additionally has some imaging parameters that can be specified by the user, such as background pixel intensity cutoffs, complexity-cost, scale, or tolerance parameters. Suppose there are five such parameters in each algorithmic set, each having a low, medium, or high value in a reasonable range. Using the Key Project system, one could analyze the microarray slide images using each of the algorithm sets and a range of parameters to obtain, say, an analysis based on each of $3 \times 3 \times 5 = 45$ combinations of imaging methods. These analyses could then be averaged and deviant analytic results investigated using statistical meta-analysis techniques that would also be built into the system. In addition, we could consider employing Bayesian statistical methods to average-out the effect of imaging-related variability from the analysis, thereby obtaining a composite estimate that does not rely on a specific imaging protocol.

In the following example, we used the MIDAS Key Project to specify an imaging envelope around a spotted array image, having the usual red and green channels. For simplicity, rectangular areas were drawn on the image to identify 100 spots, and only two imaging choices were compared. For each of the rectangles, quantitation proceeded by setting a background threshold and computing the average pixel intensity. Two different threshold values were used, 5 and 25. These background levels are virtually indistinguishable

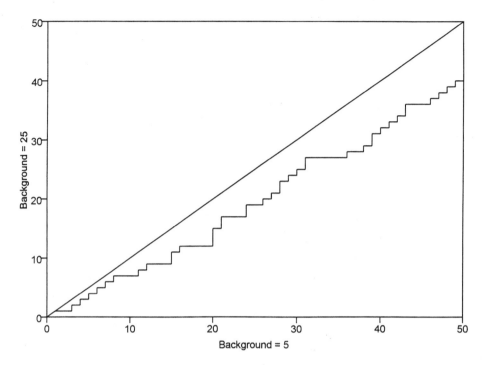

Fig. 4. Comparison of gene ranking of fold change between background levels of 5 and 25.

when visualized; visual comparison with the TIGR image suggested that either may be a reasonable choice. Thus the imaging envelope consisted of two members. The statistical analysis consisted of estimating fold change between the red and green channel and computing the corresponding rank of each gene.

Figure 4 presents a comparison of the relative ranks of the genes, across this simple imaging envelope. The height of the curve is the number of genes in the intersection of the top x ranked using a background value of 25 vs a background of 5. For example, at the value 10 on the horizontal axis, the height of the curve is 7, indicating that only 7 of the genes using a background of 25 overlap with the top 10 using a background of 5. Even in this simple example, inferences derived using two parameter values in a reasonable neighborhood demonstrate only 80% consistency. In more complicated situations, 30–60% or more additional and previously unrecognized variability may be captured in a reasonable imaging envelope. In this example, the inferences drawn across the imaging envelope could be meta-analyzed to form a consensus inference concerning the order of differentially expressed genes.

6. Conclusion

In every experimental context in which images are captured in the process of obtaining information, it is important to realize that the images are the data. Historically, inadequate attention has been paid to this viewpoint. As a researcher, one seeks conclusions that are resistant to the peculiarities of any particular imaging methods used in the process by which inference is obtained. The main benefit to an investigator is the ability to account for various factors within the imaging phase of the experiment. As detailed earlier, factors such as background signal, geometry characterization, and signal thresholding can and do have an effect on the resulting data, which in turn affects downstream analysis. Control and awareness of these influences allows an investigator to conduct inference that better reflects the underlying biology. We suggest that the MIDAS Key Project described in this chapter and the paradigm on which it is based provide such an approach, enhancing the completeness of data analysis and leading to better models for inference in image-based experiments.

3

Introduction to Microarray Experimentation and Analysis

Peter Gieser, Gregory C. Bloom, and Emmanuel N. Lazaridis

1. Introduction

Microarray experiments try to measure simultaneously the quantity of many specific messenger RNA (mRNA) sequences contained in a sample. These quantities are called *gene expression*. The sample mRNA can be extracted from human tissue, plant material, or even yeast. Because thousands of these sequences can be measured in a single experiment, scientists have a large window into the workings of a biological system. This is in contrast to use of more traditional approaches such as Northern blots, which limit research to one-gene-at-a-time experiments.

There are many ways that microarrays can be used to further scientific research. One application is in the area of human cancer where, for example, we seek to identify colon cancer patients who are at risk for metastasis. While surgical extirpation of colorectal cancer remains the primary modality for cure, patients who have metastasized to distant sites at the time of surgical intervention frequently die from their disease. Unfortunately, there is no accurate means of identifying the patients who are at risk for metastasis using current staging systems, which are based only on clinicopathologic factors. Moreover, attempts at improving these staging systems, using molecular techniques to assay the expression of single or a small number of genes, have been relatively unsuccessful. This is likely because the process of metastasis is complex and linked to the expression of numerous gene families and biological pathways. Because microarray technology provides a more comprehensive picture of gene expression, experiments involving colon cancer and metastatic tumor specimens can be used to derive a molecular fingerprint in primary tumors portending metastasis.

From: *Methods in Molecular Biology, vol. 184: Biostatistical Methods*
Edited by: S. W. Looney © Humana Press Inc., Totowa, NJ

2. Robotic Spotting vs Photolithographic Technology

The term *microarray instance* refers to a single image of a particular hybridized microarray chip, slide, or filter. Currently, there are two major technologies for generating microarray instances. The older technology uses a robotic arm to first place specific cDNA probes, representing known genes or expressed sequence tags (ESTs)[1] of interest, onto a substrate. An RNA sample is then reverse transcribed and tagged with a fluorescent dye. This mixture is washed over the substrate, where the tagged cDNA hybridizes to the complementary cDNA probe. When scanned by a laser of the appropriate wavelength, the amount of fluorescence (as seen by a confocal microscope) is a measure of the quantity of tagged cDNA that has adhered to each probe. This, in turn, is used to directly infer the amount of a particular gene present in the original sample. In addition, a second RNA sample, tagged with a different fluorescent dye, can be mixed with the first sample. By scanning at two different wavelengths, information from each sample can be generated using only one slide. A newer, proprietary, technology by Affymetrix has emerged that also generates microarray instances. The Affymetrix GeneChip system works by creating a defined array of specific oligonucleotides, or *oligos*, on a solid support via appropriate sequencing of masks and chemicals in a photolithographic process, not unlike the way in which semiconductors are manufactured. A Biotin-tagged cRNA sample is washed over the chip and hybridized to the complementary oligo probes. A laser scans the chip and the fluorescence is measured. In contrast to the spotting technology, a mathematical model is required to combine the information from multiple oligos into a single gene expression level.

These two methods are similar in that they both contain probes in an array on a solid surface and are exposed to a sample for hybridization. Both are scanned and result in an image representation of the data.

The differences in the methods are key. One difference is the manner in which the expression level for a gene is established. The Affymetrix system uses a designed set (typically 40) of 25-mer oligos per gene,[2] which must be combined to quantify gene expression. This is in contrast to the spotting technology, where each gene is typically represented by only one target sequence. Another difference is the ability of the spotted array to generate two or more microarray images from the same slide by scanning it at different frequencies, corresponding to the fluorescent labels employed.

---

[1] Successive references to genes or gene expression implicitly include ESTs.

[2] The 40 features in a gene set are typically composed of 20 perfect match–mismatch oligo pairs. Each perfect match oligo is a true cDNA probe for its associated gene product. Each mismatch feature contains a single nucleic acid substitution in the center of the strand relative to the perfect match. Affymetrix includes mismatch features on chips to provide a means for quality control prior to and during quantitation.

There are statistical implications that need to be considered in working with the different technologies. As spotted arrays have the ability to utilize multiple samples per slide, they are more flexible than the Affymetrix arrays in accommodating different types of experimental designs. However, spotted arrays can require human intervention that is hard to account for in statistical models. Supervision in the form of discarding malformed spots is common, but what are the implications? What methods are used to find the spots, and what impact does the choice of a particular method have on the final analysis? As argued in Chapter 2 by Bloom, Gieser, and Lazaridis, imaging choices made within a reasonable envelope can have substantive effects on analytic results. The Affymetrix arrays, although generally more consistent and well-defined on the substrate, use only one sample chip, and that limits the choice of experimental design. The additional complexity of trying to put together the information from the individual oligos also provides a statistical challenge that is not present with the spotted arrays.

It is an important fact that microarray technologies continue to develop, resulting in additional complexities for the analyst. For example, it has been suggested that 60-basepair oligo sequences may improve sensitivity and specificity for gene expression, relative to the 20-basepair oligos currently used in Affymetrix chips. It is also of note that fewer oligo probes for each gene set may be assembled in future versions of Affymetrix chips. Data from spotted arrays are being impacted by developments in substrate technology seeking to improve the imaging properties of glass slides. Other ongoing technological developments include the use of ink-jet spotters as a means to place probes on a substrate, and increases in the acceptable density of spots or features on a microarray instance. Changes in technology imply that experiments performed at one time may need to be treated differently from those performed at a later date. Because one goal of many microarray projects is to compile gene expression information over multiple years, there is a need for analytic models that can handle the complexities resulting from further technological developments.

3. Imaging Analysis of Microarray Images

As mentioned previously, we use the term *microarray instance* to refer to a single image of a particular hybridized microarray chip, slide, or filter. Such terminology heeds the fact that the first (electronic) capture of information from a physical experiment is in the form of an image, which is typically obtained using a laser scanning device. Because so many imaging issues can impact what quantities are derived from a microarray image, we advocate treating microarray images as "the data." Although Chapter 2 suggests a new paradigm for microarray data analysis based on this thinking, the current approach at most institutions derives only one set of data from any single microarray instance. Thus, we limit our discussion of imaging analysis of microarray images to a few, brief remarks.

The primary elements of imaging analysis that can affect quantitation are choice of background adjustment and spot characterization methods, along with choice of their associated parameters. Typical background adjustment algorithms may account for global (image-wide) and/or local (in the vicinity of a spot or feature) background phenomena. What regions of an image are chosen to calculate the parameters of these algorithms may vary substantially by technology and analyst. For example, Affymetrix technology packs oligonucleotide probes so densely on the surface of a chip that one must employ minimally hybridized probe regions to calculate average background intensity. Global background values can be calculated directly using this scheme; calculation of local background adjustments requires application of a spatial model that uses a method such as kriging. Because of the manufacturing and hybridization processes employed by oligonucleotide chip technology, local background adjustments may be relatively less important than in the context of spotted arrays, in which the application of a coating to a glass slide as well as other technological issues can introduce substantial amounts of local noise. Chapter 2 presents an example wherein use of two, virtually indistinguishable, threshold values for local background lead to substantially different inference.

Spots may also be characterized differently in different applications. Three common approaches involve quantitating image intensity in rectangular regions, in ellipsoids, or even in regions determined by an edge detection scheme (which will also depend on choice of background threshold). In addition, spotted microarrays may exhibit doughnut spotting, that is, bright spots with a dark hole in their centers resulting from the process by which probe material was placed on the substrate by a robotic needle. An algorithm to identify and discount these dark regions may be appropriate in certain cases.

4. Statistical Analysis of Microarray Data
4.1. General Overview

Having illustrated that quantities derived from microarray instances can be affected by imaging choices, we proceed to discuss the statistical issues that are the focus of this chapter, restricting our presentation to cases wherein only one set of data is derived from any single microarray instance. In such a set, each arrayed gene is associated with one estimate of its (relative) expression. We employ the generic term *objects* to mean a set of genes, microarray instances, drugs, or so on, to which an analytic method for clustering or classification can be applied.

Generally, *microarray data* consist of observations on *n*-tuples of objects. A common form for these data gives one observation of estimated gene expression for each combination of elements of a set of genes and a set of microarray instances.

An important analysis characteristic is the degree to which external information is employed to assist a method in determining appropriate object clusters or classes. Methods that rely on external information or user interaction are called *supervised*. Methods that refer only to the data at hand are called *unsupervised*. In a similar manner, some methods may be trained using external data (and in that sense are supervised), but may still be applied to new data in an unsupervised manner.

All methods assume a certain degree of structure in the data to be analyzed. Whether the underlying model is explicitly recognized or not, some methods are structurally heavy, leading to a substantial influence on clustering or classification, while structurally light methods tend to have less influence. Clearly, a good model will reflect the structure of the data, and the best models will represent the underlying biology.

4.2. Data Adjustment

Because the data represented by a single microarray instance are a reflection of the relative amounts of gene expression in a tested sample, an important question is how to standardize these quantities to allow for comparison across multiple instances. It is unfortunate that some authors call this process *normalization*, as that term additionally suggests the transformation of data to satisfy Gaussian distributional properties. We discuss data transformation for modeling purposes toward the end of this chapter. Several methods for standardizing across microarray instances are available, but each has its disadvantages.

The most basic technique involves standardizing data from each chip according to the average intensity of the pixels across the whole scan of the chip, or across all the spots. This approach has the advantage of simplicity, and requires no special experimental considerations, but is fraught with danger. For example, the goal of many experiments involves depressing or stimulating transcription of a large number of gene products, so that different overall average intensity between images may not be an imaging artifact.

Another basic technique is to standardize data from each chip according to the average intensities of internal controls, or "housekeeping" genes, that are not expected to change across a particular experiment (such as cellular gene products in a viral DNA chip). This method avoids the major problem of the previous approach, but still relies on the assumption that the internal controls are unaffected by differences in samples and experimental conditions across microarray instances. If a large quantity of internal control spots is assembled on a microarray, then the approach of Amaratunga and Cabrera *(1)* may be considered. These authors employed the intensity histograms of pixels associated with the internal controls to estimate a transformation of each microarray instance to a standard intensity curve. Alternatively, one can try dividing the average intensity measurement of each spot by the average intensity over a small set of controls. Most published analyses to date employ this approach, but this can be dangerous because the intensity transformation between two images (and especially between two fluorescent channels) is frequently nonlinear.

A third technique standardizes data according to the fluorescence of one or more known elements added to the experimental sample just prior to hybridization. Required is the assumption that the addition of a "spike" to the mixture does not change the sample's other properties. For example, the Affymetrix procedure is to spike human samples with herring sperm DNA, relying on the supposition that herring sperm and human DNAs are sufficiently lacking in homology. One major problem is the likelihood that the image intensity of a spike will demonstrate substantial variability. To account for this, the Affymetrix protocol recommends spiking the sample with a set of staggered concentrations of control cRNAs. These controls are used to determine the sensitivity of the chip by noting the smallest concentration level that can be detected. We note that they are not used in the Affymetrix analytic procedure except as a global filter for adequacy of the microarray instance.

Finally, any of a variety of statistical regression models can be used to standardize data by looking at pools of experimental samples in an associated experimental design. As an illustration of how one such approach might work, suppose there are two biological samples to be analyzed using a one-channel scan microarray system. Instead of running each sample on each chip separately, suppose the first sample is run on the first chip, and a mixture of the two samples is run on the second chip. Denoting the expression vector of each sample by X_i and the expression vector from each microarray instance by Y_j, one might (somewhat naively) expect that the above experiment would satisfy the relationships $X_1 = Y_1$, and $X_2 = 2(Y_2) - X_1$. Unfortunately, this may not always be the case, possibly owing to effects of RNA concentration and complexity in the mixture. Experiments suggest a trend correlated with spot average intensity, whereby mixtures of samples on a single chip tend to underestimate the average of samples run on different chips. However, because this relationship may be predictable, it is possible that a regression model might be used to adjust for possible bias.

4.3. Combining Oligonucleotide Information in a Probe Set

A special requirement of oligonucleotide chips is a robust method to combine the measured average intensities of oligo features in a probe set into a single value estimating expression of the associated gene product. Typically, each probe set is composed of 40 features, arranged in 20 perfect match–mismatch oligo pairs. Each perfect match oligo is a piece of cDNA probe for the associated gene product. Each mismatch feature contains a single nucleic acid substitution in the center of the strand relative to the perfect match. Affymetrix includes mismatch features on chips to provide a means for quality control prior to and during quantitation. We note that the existence of perfect match and mismatch oligos in each probe set on an Affymetrix chip is not an important component of this problem because there is substantial evidence to suggest that molecules with high affinity to a mismatch oligo may have low affinity to the corresponding perfect match. Thus, we advise against the use of algorithms that combine perfect match and mismatch information to create summary statistics intended to estimate gene expression, such as pairwise differences in average feature intensity.

The current software provided by Affymetrix to investigators returns the average of feature intensities for a subset of features in each probe set. The subset is chosen in each array by considering the mean and standard deviation of differences between paired perfect match and mismatch average intensities, after excluding the maximum and the minimum. Probes whose probe pair differences deviate by more than three standard deviations from the mean are excluded in gene expression estimates. If two microarray instances are to be compared, the intersection of acceptable probes in each instance is employed to evaluate gene expression difference. Not only does this procedure potentially exclude informative probes with large responses in individual arrays, it also implicitly assumes that the information provided by each acceptable oligo is of equal importance. Clearly superior to this approach would be a weighted sum of oligo-specific values, with parameters chosen to reflect the extent of information in each oligo.

At least two procedures have been employed to choose such weights. Li and Wong *(2)* assume that the average intensity of each probe in a probe set increases linearly with respect to increases in underlying, unknown gene expression, but with probe-specific sensitivity. This assumption leads to a weighted sum conditional least squares estimate of gene expression. In what follows we ignore their use of mismatch feature intensities. Letting i equal index array instances, j index probes, and n index genes, their basic model is $y_{ij} = \theta_i \phi_j + \varepsilon_{ij}$ with $\varepsilon_{ij} \sim N(0, \sigma^2)$, and $\sum_j \phi_j^2 = J$, J being the number of probes in a given probe set. Least squares estimation may be performed by iteratively calculating the gene-specific parameters (θ_i) and the probe-specific weights (ϕ_j), identifying and excluding during the iterative procedure any outlier microarray instances and probes (outliers relative to the model) as well as probes with high leverage (which may be untrustworthy because of their influence on the model estimates). Possible drawbacks to this approach arise from reliance on the parametric model, on its distributional assumptions, and on the criteria one employs to exclude outlying or untrustworthy data.

In our setting we employ a nonparametric approach to weighing oligo features using a minimum risk criterion, an approach that is easily described and implemented. We think of oligos in a probe set as players on a team, who have been selected on evidence that as individuals they are top performers. The performance of any given player can be gauged according to how well that player estimates a value of interest, in this case, the expression of the gene associated with the probe set. Each microarray instance in a particular analysis corresponds to a single game, resulting in a score for each of the members of the team. Indexing players by i and games by j, denote the average intensity of each oligo at each microarray instance by y_{ij}. Our main question involves a coaching decision, whereby the analyst (coach) seeks to obtain a better estimate of (unobserved) gene expression, θ, for a particular probe set. Using the least squares criterion, it can be shown that a coach should use $\bar{y}_{gj}$ to estimate θ in the situation where all the players perform equally well across games. In the microarray context, relative performance of players must be estimated by the coach, who does not know the value of θ that would be needed to calculate an exact loss function. Instead, we argue that within each game, the rational coach would evaluate each player against a best estimate of gene expression derived from the rest of the team. Thus, the problem of minimizing loss over players and games reduces to calculating a set of parameters, ϕ_i, such that

$$\sum_i \sum_j \left(\phi_i y_{ij} - \frac{\sum_{k:k \neq i} \phi_k y_{kj}}{\phi_g - \phi_i} \right)^2$$

is minimized. The fact that this procedure is equivalent to minimizing the leave-one-out cross-validation estimate of variance for the mean of coach-adjusted player estimates, $\phi_i y_{ij}$, suggests the situations in which this approach may perform substantially better than a parametric one, including settings in which limited information is available about probe weights. Designed laboratory experiments, in particular, typically result in few microarray instances over a multiplicity of conditions. In addition, situations in which assumptions of a linear model with constant variance may be violated will often arise in laboratory experiments because of the hybridization performance of different molecular mixtures across samples being compared.

4.4. Differential Gene Expression

After proper standardization, a natural question to ask is which genes are differentially expressed across two or more samples. For concreteness, consider a spotted microarray in which two samples have been cohybridized.

Traditionally, the raw ratio (between the two samples) of standardized spot average intensities has been used to make inferences about which genes are significantly differentially expressed. Newton et al. *(3)* point out that this is problematic because a given fold change may have a different interpretation for a gene whose absolute expression is low in both samples as compared to a gene whose absolute expression is high in both samples. They suggest that there is room for improvement in the initial signal processing that may have bearing on downstream tasks such as clustering.

The solution Newton describes is based on hierarchical models of measured expression levels that account for two sources of variation. One source is measurement error, the fluctuation of the spot intensity around some mean value that is a property of the cell type, the particular gene, and other factors. The second source is gene variation, the fluctuation of the mean intensity value between the different genes. This formulation allows the computation of probabilistic statements about actual differential expression. The key findings are that observed ratios are not optimal estimators, focusing on fold changes alone is insufficient, and confidence statements about differential expression depend on transcript abundance.

The specific sampling model used by Newton is based on the Gamma distribution. Given genes are modeled as independent samples from distinct Gamma distributions with common coefficient of variation (i.e., constant shape parameter). Specifically, if R and G are the measured expression levels for a gene across the two samples, let $R \sim \text{Gamma}(a, \theta_R)$ and $G \sim \text{Gamma}(a, \theta_G)$. Then the scale parameters are assumed to follow a common $\text{Gamma}(a_0, \upsilon)$ distribution. This model is stated to be reasonably flexible and skewed right, while exhibiting increasing variation with increasing mean. It turns out that given these model components, the Bayes estimate of differential expression is $p_B = (R + \upsilon)/(G + \upsilon)$, which has the classic form of a shrinkage estimator. The implication of this is that for strong signals p_B will be close to the naïve estimator R/G, but there is attenuation of p_B when the overall signal intensity is low. Clearly, p_B naturally accounts for decreased variation in differential expression with increasing signal on the log scale. One problem with this method, however, is that spots that cannot be distinguished from the background in either channel are omitted from analysis. It could be argued that these are in fact the most important cases of all! Another problem is that the restrictive parametric model may not fit the distribution of actual gene expressions on a given chip.

To determine significant differential expression, an additional layer is added to the model in the form of a latent variable z that indicates whether or not true differential expression exists. The Expectation-Maximization (EM) algorithm is then used to estimate the parameters and compute the posterior odds of change at each spot.

An alternative method for exploring differential expression is via *robust* analysis of variance (ANOVA). This has been described by Amaratunga and Cabrera *(1)*.

4.5. Principal Components

Although principal components analysis (PCA) is not a model-based method, it still plays an important role in facilitating model-based analyses. PCA is a technique commonly used for dimension reduction. Generally, PCA seeks to represent n correlated random variables by a reduced set of d ($d < n$) uncorrelated variables, which are obtained by transformation of the original set onto an appropriate subspace. The uncorrelated variables are chosen to be good linear combinations of the original variables, in terms of explaining maximal variance, orthogonal directions in the data. Data modeling and pattern recognition are often better able to work on the reduced form, which is also more efficient for storage and transmission. In particular, pairs of principal components are often plotted together to assist in visualizing the structure of high-dimensional data sets, as in the biplot.

Suppose we have a set of microarray data in standard form, X, a matrix with as many rows as there are genes (M responses) and as many columns as there are microarray instances (n observations). Standard PCA seeks the eigenvectors and associated eigenvalues of the covariance matrix for these data. Specifically, if Σ is the covariance matrix associated with the random vector $Z = (Z_1, Z_2,..., Z_n)$ and Σ has eigenvalue–eigenvector pairs $(\lambda_1, \overset{v}{e}_1), (\lambda_2, \overset{v}{e}_2), ...,(\lambda_n, \overset{v}{e}_n)$ where $\lambda_1 \geq \lambda_2 \geq \cdots \geq \lambda_n \geq 0$, then the ith principal component is given by $Y_i = e_{1i}Z_1 + e_{2i}Z_2 + \cdots + e_{ni}Z_n$. Note that the principal components are uncorrelated and have variances equal to the eigenvalues of Σ. Furthermore, the total population variance is equal to the sum of the eigenvalues, so that $\lambda_k/(\lambda_1 + \cdots + \lambda_n)$ is the proportion of total population variance due to the kth principal component. Hence if most of the total population variance can be attributed to the first one, two, or three components, then these components can "replace" the original n variables without much loss of information. Each component of the coefficient vector also contains information. The magnitude of e_{ki} measures the importance of the kth variable to the ith principal component, irrespective of the other variables. In the context of microarrays, the lack of data in the instance direction deemphasizes the data-reduction aspect of principal components. Instead, the interest is generally in the interpretation of the components.

The sample principal components are calculated as described earlier, replacing the (generally unknown) population covariance matrix with the sample covariance matrix. Hilsenbeck et al. *(4)* applied this technique to three microarray instances generated from human breast cell tumors. These instances correspond to estrogen-stimulated, tamoxifen-sensitive, and tamoxifen-resistant growth periods. Their results yielded three principal components interpreted as (1) the average level of gene expression, (2) the difference between estrogen-stimulated gene expression and the average of tamoxifen-sensitive and tamoxifen-resistant gene expression, and (3) the difference between tamoxifen-sensitive and tamoxifen-resistant gene expression.

Another use of principal components is as a basis for clustering. The correlation of each gene with the leading principal component provides a way of sorting (or clustering) the genes. Raychaudhuri et al. *(5)* analyzed yeast sporulation data, which measured gene expression at seven time points *(6)*. They determined that much of the observed variability can be summarized in just two components: (1) overall induction level and (2) change in induction level over time. Then they calculated the clusters according to the first principal component and compared them to the clusters reported in the original paper.

It is also possible to use PCA to reduce the dimensionality of the analytic problem with respect to the gene space. A number of ways have been proposed to do this.

One use of this idea is in *gene shaving (7)*. This method seeks a set of approximate principal components that are defined to be *supergenes*. The genes having lowest correlation with the first supergene are shaved (removed) from the data and the remaining supergenes are recomputed. Gene blocks are shaved until a certain cost–benefit ratio is achieved. This process defines a sequence of blocks with genes that are similar to one another. A major problem with this approach is the shifting definition of supergenes over the course of the analysis. Although gene shaving incorporates the ideas of principal components, it is important to recognize that the shaving algorithm itself is *ad hoc*.

Another application of the gene space reduction idea solves the problem of using gene expression values as predictors in a regression setting. Because correlated predictors are known to cause difficulties, principal components regression (PCR) uses the gene principal components as predictor surrogates. A second method that uses this idea is partial least squares regression (PLSR). PLSR is employed to extract *only* the components (sometimes called *factors*) that are directly relevant to both the predictors and the response. These are chosen in decreasing order of relevance to the prediction problem.

Both PCR and PLSR produce factor scores as linear combinations of the original predictor variables, so that there is no correlation among the factor score variables used in the predictive regression model. For example, suppose we have a data set with response variable Y and a large number of highly correlated gene expression predictor variables X. A regression using factor extraction for these types of data computes the factor score matrix $T = XW$ for an appropriate weight matrix W, and then considers the linear regression model $Y = TQ + \varepsilon$, where Q is a matrix of regression coefficients (loadings) for T, and

ε is the error term. Once the loadings, Q, are computed, the preceding regression model is equivalent to $Y = XB + \varepsilon$, where $B = WQ$, which can be used as a predictive regression model for gene expression data on the original scale.

PCR and PLSR differ in the methods used in extracting factor scores. In short, PCR produces the weight matrix W reflecting the covariance structure between the predictor variables, while PLSR produces the weight matrix W reflecting the covariance structure between the predictor and response variables.

Partial least squares regression produces a weight matrix W for X such that $T = XW$. Thus, the columns of W are weight vectors for the X columns producing the corresponding factor score matrix T. The weights are computed so that each of them maximizes the covariance between the response and the corresponding factor scores. Ordinary least squares procedures for the regression of Y on T may then be performed to produce Q, the loadings for Y. Thus, X is broken into two parts, $X = TP + F$, where the factor loading matrix P gives the factor model and F represents the unexplained remainder.

Whether centering and scaling of the data (normally done when determining principal components) makes sense for microarray data is an open question, and all the caveats that go along with PCA are still in effect *(8)*.

A Bayesian application, using a related idea, models a binary response using a probit model and decomposes the linear predictor vector using the SVD *(9)*. Specifically, if z_i is the binary variable reflecting status for each patient, then let $\Pr(z_i = 1 | \beta) = \Phi(x_i'\beta)$. Using the decomposition, obtain $X'\beta = (F'D)\gamma$, where $X = ADF$ from the SVD; A is the SVD loadings matrix; F is the SVD orthogonal factor score matrix (as before); D is the diagonal matrix of singular values; and $\gamma = A'\beta$ is the vector of parameters on the subspace formed using the new linear basis. A key part of this method is determining reasonable prior distributions for the vector of parameters on the subspace formed using the new linear basis. A reasonability criterion is defined in terms of the interpretability of the priors when back-transformed to the original space.

4.6. Latent Class Models

There are two major types of analyses that fall broadly into this category. A clustering method entails placing objects that are close together into clusters according to a specified metric. A classification method is where clustering is performed by estimating the probability of each object's membership in a latent (i.e., unobservable) class. All classification methods can be used to generate clusters, but clustering methods do not imply a particular class definition. Both clustering and classification methods can be used to discriminate among estimated differentiated sets of objects. However, because clustering methods do not explicitly model theoretical class constructs, they provide no basis for determining misclassification, the association of an object with a class to which it does not belong. This is a major disadvantage of such analyses.

Analytic methods can be geared toward clustering or discriminating among various classes of either genes or microarray instances. These are one-way analyses. Methods can also be geared toward jointly clustering or discriminating among both gene and microarray instances. These are two-way analyses.

Generally, clustering methods use similarity measures to associate similar objects and to disassociate sets of similar objects from each other. Tibshirani et al. *(10)* review the various methods of clustering and show how they can be used to order both the genes and microarray instances from a set of microarray experiments. They discuss techniques such as hierarchical clustering, *K*-means clustering, and block clustering. Dudoit et al. *(11)* compare the performance of different discrimination methods for the classification of tumors based on gene expression data. These methods include nearest-neighbor classifiers, linear discriminant analysis, and classification trees.

Although these *ad hoc* analyses are still popular owing to the current dearth of more-sophisticated techniques and software, statisticians are feverishly working to come up with model-based approaches so that inference will be possible. It is these model-based approaches that are the focus of this chapter.

Classic multidimensional latent class models specify the form of the class conditional densities. A common specification associates each class with a multivariate normal distribution *(11)*. In this case, the maximum likelihood discriminant rule is

$$C(x) = \text{argmin}_k \{(x - \mu_k)\Sigma_k^{-1}(x - \mu_k)' + \log |\Sigma_k|\}$$

Three special cases of interest are (1) when the class densities have the same covariance matrix, $\Sigma_k = \Sigma$, the discriminant rule is based on the square of the Mahalanobis distance and is linear:

$$C(x) = \text{argmin}_k (x - \mu_k)\Sigma_k^{-1}(x - \mu_k)'$$

(2) when the class densities have diagonal covariance matrices, $\Delta_k = \text{diag}(\sigma^2_{k1},...,\sigma^2_{kp})$, the discriminant rule is given by additive quadratic contributions from each variable:

$$C(x) = \text{argmin}_k \sum_{j=1}^{p} \left(\frac{x_j - \mu_{kj}}{\sigma^2_{kj}} - \log(\sigma^2_{kj}) \right)$$

and (3) when the class densities have the same diagonal covariance matrix, $\Delta = \text{diag}(\sigma^2_1,...,\sigma^2_p)$, the discriminant rule is linear:

$$C(x) = \text{argmin}_k \sum_{j=1}^{p} \left(\frac{x_j - \mu_{kj}}{\sigma^2_{kj}} \right)$$

For repeated measurement experiments, Skene and White *(12)* describe a flexible latent class model that also assumes normality. In the context of microarray experiments, let Y_{mg} denote the log-transformed average spot intensity response of gene g on microarray slide m. Let L represent a discrete latent variable with levels $1,...,J$. Assume that corresponding to each level of L is a profile of gene expression defined across multiple slides, $p_L = \{p_{1L},...,p_{ML}\}$ which is defined in terms of deviation from average gene expression. Assume also that genes in the same biological pathway may have different expression intensities, d_g, depending on such factors as gene copy number, transcription efficiency, and so on. Thus, it makes sense to let $\mu_{mgj} = a + d_g + p_{mj}$ be a model for the mean response in microarray instance m for gene g in latent class j. Conditional on latent class membership, the error is assumed normal so that $Y_{mg} \mid L = j \sim N(\mu_{mgj}, \sigma^2)$. The difficulty with this formulation is that estimation is a problem when the number of parameters is large, which is frequently the case with microarray data. Research is currently ongoing into techniques that might overcome this limitation.

Similar latent class model forms can be considered in a Bayesian context by placing Dirichlet priors on the class membership probabilities and appropriate conjugate priors elsewhere, then conducting Markov Chain Monte Carlo (MCMC) to generate samples from the full posterior to estimate class and gene parameters. The approximate EM solution is used as the starting point for the MCMC algorithm. An extension of this idea is to simultaneously divide the genes into classes with substantial internal correlation as well as allocate microarray instances to latent sample classes. Such models seek to identify gene classes with high sample class discriminatory power. Typically, we transform data to the log scale prior to application of the model.

For example, consider the following analysis of a time-course experiment on fibroblasts (13). In this experiment, cells were first serum-deprived and then stimulated, to investigate growth-related changes in RNA products over time. Other aliquots were additionally treated with cycloheximide. Samples of untreated and treated cells were collected at 12 and 4 time points, respectively, as were samples of unsynchronized cells. Microarrays included 8613 gene products, but analyses included only 517 of these. In published work, the authors used a hierarchical clustering algorithm to identify 10 patterns of gene expression in a subset of 517 genes, which was filtered before application of the clustering algorithm on the basis of the existence of "significant" univariate observed variability fold changes in gene expression over time.

Using a Bayesian latent class model of the above form, we analyzed the complete set of 8613 gene products, over all the time points and experimental conditions. We employed a normal error model on the log-transformed data with mean conditional on latent gene class and time (plus experimental condition), with gene-specific intensity modifiers to represent the degree to which each gene is a good marker of its associated latent gene class. Because genes

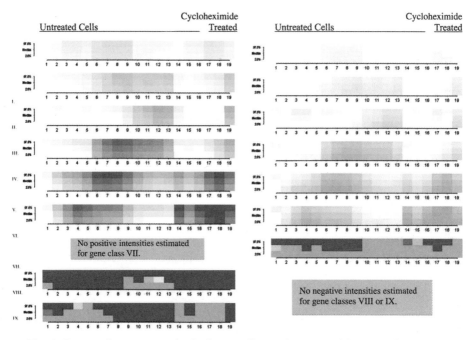

Fig. 1. Latent class patterns including median estimates and 95% confidence intervals from the analysis of serum-stimulated fibroblasts. This figure demonstrates both time-dependent increases and decreases in gene expression, as represented by positive (mostly red) and negative expression (mostly green) patterns.

could be estimated to have decreases in expression (negative intensity modification parameters) or increases in expression (positive intensity modification parameters), in **Fig. 1** we present both the estimated patterns and their inverses. In other words, red and green colors represent, respectively, higher or lower levels of expression relative to untreated cells at time 0. We use a white background to aid in visualization; brighter colors represent greater relative deviation from baseline.

This analysis also adjusted for data quality issues. Specifically, we treated an observation (one spot on one microarray instance) as missing whenever the interpixel correlation between the two scans was <0.6. Thus, our analysis accounts for and is unbiased by differential hybridization of samples. A consequence of this is that pattern VIII is separated from pattern VI because of insufficient information at samples 1 and samples 9 through 13, as evidenced by a wide confidence interval ranging from very bright green (2.5th percentile) to very bright red (97.5th percentile). Probabilities of gene membership in these latent patterns for every gene in the data are estimated by the model.

This analysis illustrates some additional advantages of the latent class approach. First, estimated time-course patterns are smoother than those originally proposed by hierarchical clustering analysis, even though no smoothness criterion is imposed by the model. Although no assumptions were made regarding the correlations within treatment group across time, the estimates show expression patterns that are reasonably correlated with known stages of cellular growth and mitosis, suggesting that this model is uncovering the underlying biology. Second, no prefiltering of genes is required by the technique; prefiltering is often necessary to apply clustering methods. Third, the model adjusts for data quality issues using Bayesian statistical approaches, to reduce the potential biases that can be introduced by experimental variability, especially at low spot intensities.

Standard statistical approaches can also be employed to diagnose lack-of-fit of this model to the data. For example, residual plots can be employed to identify deviations from the model fit. Residual distances between the observed and predicted values, conditioning on the Gaussian model, follow a chi-square (χ^2) distribution in analogous fashion to the residuals from standard multivariate analyses *(14)*. In that context,

$$n\left(\overline{X} - \mu\right)^T \Sigma^{-1}\left(\overline{X} - \mu\right) \sim \chi^2\left(p\right)$$

where $(\overline{X} - \mu)$ is the vector of differences between observed and expected mean values, Σ^{-1} is the generalized inverse of the covariance matrix, and $\chi^2(p)$ refers to the χ^2 distribution on p degrees of freedom, p being functionally related to the dimensionality of the data. We calculated values of the form $\left(\overline{X} - \mu\right)^T \Sigma^{-1}\left(\overline{X} - \mu\right)$ and compared their quantiles to those of the appropriate χ^2 distribution in **Fig. 2**. Points falling on the straight line give evidence that the model fits the signal in the data, because the residuals seem to follow the appropriate χ^2 distribution. Points deviating to the right of the line signify overdispersion relative to the residual χ^2, meaning that there may be some statistically significant additional signal in those gene products that is not being described in the current model. Only 18 outliers (3.4%) from this model were identified. The most discrepant gene, *AA058863*, is a Soares retina N2b4HR *Homo sapiens* cDNA clone containing ALU sequences, which may account for observed expression pattern differences. Another set of gene products was identified through the residual analysis that could be associated with a still unidentified pathway. Another set of gene products, including WEE1-like protein kinase, were underexpressed around times 7 and 8 to a much greater degree than would be expected by the model. Whereas the hierarchical clustering analysis groups WEE1 with expressed genes such as *p57Kip2* and *p27Kip1*, the latent class analysis identified differences between the cyclin-dependent kinase inhibi-

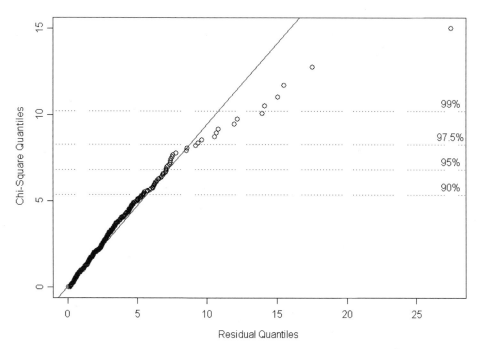

Fig. 2. QQ-plot of residuals from exploratory latent class model vs χ^2 distribution, showing outlying expressed genes deviating from the line towards upper right.

tors and the mitotic inhibitor WEE1, suggesting that WEE1 may have a slightly different function. Changes based on such observations can be fed back into the analytic model.

As previously discussed, a key distinction between formal latent class models and clustering algorithms is that the former provide an inferential framework while the latter do not. An inferential framework allows one to make probability statements concerning the uncertainty associated with identified classes and their members, as well as to estimate the number of classes needed to describe a particular data set. In the clustering arena, an approach for determining the validity of a particular clustering has recently been developed and is described in Bittner et al. *(15)*. The method is based on evaluating cluster membership after introducing random perturbations to the data set. Hierarchical clustering is performed on the perturbed data set and compared to the original tree. Comparisons involve cutting the original and perturbed trees into k clusters followed by computing the proportion of paired samples clustering together in the original tree that do not cluster together in the perturbed tree. The average over multiple perturbed data sets (for a given k) yields the weighted average discrepant pairs statistic (WADP$_k$), which is then plotted vs k. Local minima on the WADP$_k$ curve indicate reproducible levels of structure.

4.7. Differential Equations

In this subheading, we discuss what are perhaps the most structural of microarray data modeling approaches. Various authors are seeking to model genetic networks using sets of nonlinear differential equations. Examples can be found in Reinitz and Sharp *(16)* and Wahde and Hertz *(17)*. Parametric forms are employed to model the rates of change of expression in certain genes. Unfortunately, lack of data restricts the estimates to first-order terms in the differential equations, so that only gene interaction weights can be estimated. Two assumptions in the Wahde and Hertz *(17)* model are of note. The first is in the choice of a particular structure for the combination of gene effects (the activation function; a genetic algorithm was employed for estimating the parameters), which assumes linear transcription. The second is the use of average trajectories of coarse clusters of genes with similar expression patterns as nodes. For these authors, this simplification is inspired by the fact that for reliable determination of the network parameters, a minimum requirement is that the number of useful data points exceeds the number of parameters. Gene expression data series often consist of only a few measurements, and clustering of genes into sets with similar temporal expression patterns substantially reduces the number of parameters requiring calculation.

Another example is in Chen et al. *(18)*. These authors also propose a differential equation model for gene expression, and provide a method to construct the model from temporal expression data. They make a number of assumptions, among which are a linear transcription function for each gene and feedback of the gene translation product on the transcription rates. They discern in their model transcription, translation, and degradation of RNA and proteins. Parameters can be estimated for these processes using Fourier transforms for stable systems, an approach that is specific for genes with periodic expression. These are important in cell cycle studies, for example, and all genes considered in the model are assumed to show this kind of expression pattern. As before, reduction of the problem, this time by employing periodicity and requiring stability in their system, is essential to solving this system with today's technology. The reduction in complexity resulting from these assumptions is substantial enough that many more features of gene expression could be parametrized and estimated than in the Wahde and Hertz approach.

D'Haeseleer et al. *(19)* provide a third example. These authors begin analysis by calculating cubic spline smoothers, fitting the curves of the gene expression time-series data. Of note is the fact that they use data from three different experiments that employed different time scales. Using a first-order interaction matrix as in Wahde and Hertz *(17)*, they estimate the interaction parameters using a least squares fit to the smooths. Limitations of this approach include a lack of reducibility in the gene interaction structure and restriction to the primary linear components of the system.

Clearly, this kind of approach is promising, but must undergo substantial development before it can become widely applicable in data analysis. Significant research is needed on the interface of applied mathematics and statistics, so that known physical functioning of biological pathways can be appropriately reflected in otherwise data-driven statistical models.

4.8. Additional Issues

In this subheading we address three questions frequently raised regarding microarray data analysis. We consider these issues after describing the bulk of methods in the preceding subheadings, primarily because answers to them depend on what downstream analytic techniques one seeks to apply.

The first concerns whether microarray data should be transformed to a scale other than the one in which they were collected. The answer is simple: it depends on what analysis one seeks to perform. Although there are good biological reasons to consider transforming data to a $\log_{10}$ scale, we have found no situations in which distributional assumptions in a statistical model could be guaranteed through data transformation. In part this is a reflection of Kolmogorov's lament concerning the perpetual lack of fit one observes in employing simple forms to model large data sets. Our advice to end-user analysts is to pick methods that are relatively robust to a reasonable set of data transformations. We have had better luck with latent class models in this regard than with principal components and related dimension reduction techniques, which are widely known to be sensitive to choice of scale.

A second question concerns whether it matters which microarray instance is used as a reference in standardizing several for analysis. Again the answer depends on what downstream analysis one seeks to do. In our work, we prefer that standardization occur synchronously with the actual data analysis, by the same model that will answer a biological question of interest. A consequence of this preference is that we find ourselves restandardizing sets of data multiple times in multiple ways over the course of an analysis. Of course, methods that are more inferentially robust in the context of data transformations are also more robust to peculiarities of standardization.

Finally, we are often asked how one should visualize microarray data. The answer to this question has two parts. First, there is no biological question of interest that can be answered by looking at a microarray image, unless of course the array was specifically designed to do so. Usually, consideration of colorful array images is worthwhile only if one suspects quality control issues resulting from biological experimentation or imaging analysis. Second, there are at least as many reasonable ways to visualize these data as there are inferential methods. Some methods (such as principal components and its relatives such as multidimensional scaling) are designed to reduce data to a small number of reasonably informative 2- or 3-D scatterplots. Other methods (clustering and latent class analysis) suggest graphical methods to visualize pattern sets. We also employ the full standard statistical repertoire of diagnostic plots in our work. In summary, there is nothing particularly special about microarray data that requires a different treatment from other large data problems with respect to visual presentation.

5. Conclusion

The vast amount of data generated by microarray technology tests statisticians' abilities to extract meaningful information from any given experimental context. The intense interest in and great potential of high-throughput, comprehensive molecular biology technologies is fueling a corresponding surge in statistical research on analytic methods for large, complex data sets. In addition to statisticians, molecular biologists, computer scientists, and imaging scientists all have roles to play in this development. Because of the multidisciplinary nature of this field, one of the greatest challenges to statisticians is in making sophisticated statistical methods accessible to their collaborators. This chapter has sought to give an overview of the most promising approaches for analyzing microarray data.

References

1. Amaratunga, D. and Cabrera, J. (2000) *Analysis of data from viral DNA microchips.* Technical Report, Rutgers University.
2. Li, C. and Wong, W. H. (2001) Model-based analysis of oligonucleotide arrays: Expression index computation and outlier detection. *Proc. Natl. Acad. Sci. USA* **98,** 31–36.
3. Newton, M., Kendziorski, C., Richmond, C., Blattner, F., and Tsui, K. (1999) *On differential variability of expression ratios: improving statistical inference about gene expression changes from microarray data.* Technical Report, University of Wisconsin.
4. Hilsenbeck, S., Friedrichs, W., Schiff, R., O'Connell, P., Hansen, R., Osborne, C., and Fuqua, S. (1999) Statistical analysis of array expression data as applied to the problem of Tamoxifen resistance. *JNCI* **91,** 453–459.
5. Raychaudhuri, S., Stuart, J., and Altman, R. (2000) Principal components analysis to summarize microarray experiments: application to sporulation time series. *Pacific Symposium on Biocomputing* Jan. 4–9, 2000, Hawaii, pp. 452–463.
6. Chu, S., De Rivi, J., Eisen, M., Mulholland, J., Botstein, D., Brown, P., et al. (1998) The transcriptional program of sporulation in budding yeast. *Science* **282,** 699–705.

7. Hastie, T., Tibshirani, R., Eisen, M., Brown, P., Ross, D., Scherf, U., et al. (2000) *Gene shaving: a new class of clustering methods for expression arrays.* Technical Report, Stanford University.

8. Hadi, A. and Ling, R. (1998) Some cautionary notes on the use of principal components regression. *Am. Statist.* **52,** 15–19.

9. West, M., Nevins, J., Marks, J., Spang, R., and Zuzan, H. (2000) *Bayesian regression analysis in the "Large p, Small n" paradigm with application in DNA microarray studies.* Technical Report, Duke University.

10. Tibshirani, R., Hastie, T. Eisen, M., Ross, D., Botstein, D., and Brown, P. (1999) *Clustering methods for the analysis of DNA microarray data.* Technical Report, Stanford University.

11. Dudoit, S., Fridlyand, J., and Speed, T. (2000) *Comparison of discrimination methods for the classification of tumors using gene expression data.* Technical Report, University of California, Berkeley.

12. Skene, A. and White, S. (1992) A latent class model for repeated measurements experiments. *Statist. Med.* **11,** 2111–2122.

13. Iyer, V. R., Eisen, M. B., Ross, D. T., Schuler, G., Moore, T., Lee, J. C., et al.(1999) The transcriptional program in the response of human fibroblasts to serum [see Comments]. *Science* **283,** 83–87.

14. Johnson, R. A. and Wichern, D. W. (1992) *Applied Multivariate Analysis,* 3rd ed. Prentice Hall, Englewood Cliffs, NJ.

15. Bittner, M., Meltzer, P., Chen, Y., Jiang, Y., Seftor, E., Hendrix, M., et al. (2000) Molecular classification of cutaneous malignant melanoma by gene expression profiling. *Nature* **406,** 536–540.

16. Reinitz, J. and Sharp, D. (1995)Mechanism of eve stripe formation. *Mech. Dev.* **49,** 133–158.

17. Wahde, M. and Hertz, J. (1999) *Coarse-grained reversed engineering of genetic regulatory networks.* Technical Report, Nordic Institute for Theoretical Physics.

18. Chen, T., He, H., and Church, G. (1999) Modeling gene expression with differential equations. *Pacific Symposium on Biocomputing* 1999, **4,** 29–40.

19. D'Haeseleer, P., Wen, X., Fuhrman, S., and Somogyi, R. (1999) Linear modeling of mRNA expression levels during CNS development and injury. *Pacific Symposium on Biocomputing* Jan. 4–9, 1999, Hawaii **4,** 41–52.

4

Statistical Methods for Proteomics

Françoise Seillier-Moiseiwitsch, Donald C. Trost, and Julian Moiseiwitsch

1. Introduction

What is Proteomics? The term *proteome* denotes the PROTEin complement expressed by a genOME or tissue. While the genome is an invariant feature of an organism, the proteome depends on its developmental stage, the tissue considered, and environmental/experimental conditions. There are more proteins in a proteome than genes in genome (which is particularly true for eukaryotes). For instance, there are several ways to splice a gene to generate messenger ribonucleic acid (mRNA). Furthermore, proteins can undergo posttranslational alterations such as truncation at the amino- (N)- and carboxy (C)-terminus and addition of saccharide or phosphate groups.

Two-dimensional polyacrylamide gel electrophoresis (2D-PAGE) is currently the only method able to separate thousands of proteins. Mammalian cell samples, for example, exhibit more than 2000 proteins (**Fig. 1**). Two coordinates characterize each protein: its isoelectric point and its molecular mass.

For the first dimension, proteins are focused electrophoretically along a pH gradient. Their movement stops when they reach a position at which they have no net charge (i.e., their isoelectric point). For the second dimension, proteins are soaked in sodium dodecyl sulfate (SDS) so that all proteins acquire the same charge density. They are then separated orthogonally by electrophoresis on a polyacrylamide gel according to their molecular weight. Under carefully controlled experimental conditions, these two dimensions, the isoelectric point and molecular mass, are independent. The separated proteins are then stained with fluorescent dyes so that they are readily detectable. The image of the displayed proteins defines the proteome. This image is digitally scanned into a database for storage (*1–4*).

From: *Methods in Molecular Biology, vol. 184: Biostatistical Methods*
Edited by: S. W. Looney © Humana Press Inc., Totowa, NJ

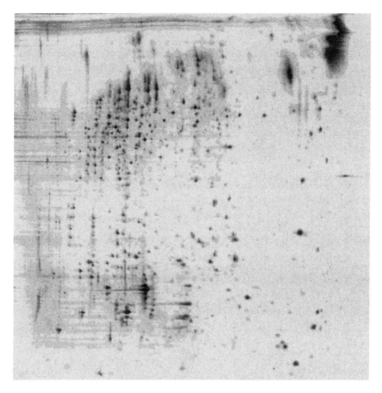

Fig. 1. 2D-PAGE image of kidney tissue sample.

This technology still presents many challenges to the experimenter. Gel quality must remain constant from day to day. Gel reproducibility from laboratory to laboratory cannot always be guaranteed. Membrane proteins necessitate a special protocol, otherwise they are underrepresented: losses are due to hydrophobic interactions between these proteins and the gel. The experimental procedure must ensure the removal of nucleic acids as these can cause streaks and artifactual migration of some proteins. Finally, the abundance threshold for detection on a gel is still to be determined.

2D-PAGE allows the systematic analysis of proteins for any disease and in any biological system. As proteins are responsible for phenotypes, they are the direct targets for therapeutic agents. Therefore, this technology has great potential for aiding in the drug-discovery process *(5)* and in medical diagnosis. The specific areas of applications are:

1. Treatment monitoring. 2D-PAGE has been used to assess treatment effects on tumors and to study overall protein expression following hormone therapy.
2. Identification of disease-specific proteins. In cancer studies, 2D-PAGE is utilized to compare protein expression in normal and cancerous tissues, therefore identifying candidate targets for drugs. It thus serves a purpose similar to that of

genomic microarray analysis where the goal is to identify clusters of disease-specific mRNAs. The disadvantage of the latter approach is that levels of a specific mRNA and the corresponding protein may not be correlated.

3. Target validation and signal transduction studies.
4. Drug mode-of-action studies.
5. Drug toxicology studies.

2. Technology Background

2.1. Electrophoresis

Electrophoresis is the process of separating a mixture of electrically charged molecules by applying an electric field. This charge is either due to charged groups on the molecules themselves (*see* **Subheading 2.1.1.**) or associated with a coating of charged molecules (*see* **Subheading 2.1.2.**). Electrophoresis is commonly used in clinical chemistry to separate macromolecules, such as proteins, in tissue samples and bodily fluids for identification or quantification. A brief overview of the physics of electrophoresis *(6)* is presented here to link the technology with the analytical methods.

Molecules can be separated via electrophoresis according to either their innate charge or their molecular weight, or both as in two-dimensional gels. Migration is also influenced by the shape of the molecule, the strength of the electric field, the ionic properties and pH of the electrophoresis buffer, and the temperature of the system. The electric field acts on the molecule according to Coulomb's law: the force on the object is proportional to its net charge (z) and the strength of the field (E). The velocity (v) of the molecule is equal to the force divided by the strength of the electric field. The potential for movement within the gel, or electrophoretic mobility (M), is then defined as v/E. With the assumption that the molecule is spherical,

$$M = \frac{ze}{6\pi\eta r}$$

where e is the electrostatic constant, η the medium viscosity, and r the macromolecular radius.

Several types of electrophoresis have been utilized since Tiselius *(7)* developed the first method, moving-boundary electrophoresis, to study proteins. Moving-boundary electrophoresis has largely been replaced by zonal electrophoresis *(8)*. This method uses a thin band, or zone, of macromolecule solution placed in a semisolid matrix such as a gel. After exposure to the electric field, the sample separates into bands of molecules with similar electrophoretic mobility. Usually gels are composed of agarose or polyacrylamide in varying percentages. However, starch and cellulose acetate have also been used and continue to have their applications. Some gels also act as a molecular sieve that separates molecules based on size.

When the electrophoresis medium comes into contact with water, hydroxyl ions adsorb to the surface, creating a stationary layer of negative charge. This layer attracts a cloud of positive ions. When the current is applied, the flow of positive ions can slow protein movement or reverse its direction. This unwanted change in mobility is called *endosmosis*. This effect tends to be strong in cellulose paper, cellulose acetate, and agarose gel, but is minimal in starch gel and polyacrylamide gel. For this reason, polyacrylamide provides better resolution, and is used when proteins of similar molecular weight need to be differentiated.

2.1.1. Isoelectric Focusing

Separating molecules by their inherent charge using electrophoresis is termed *isoelectric focusing*. In isoelectric focusing, the electrophoresis medium is constructed so that a pH gradient exists between the electrodes. Proteins have many ionization sites that are pH dependent. Such a molecule, that can be either positively or negatively charged, is called a *zwitterion*. The gel buffer must have the ability to both donate and accept hydrogen ions, which allows the charge of a zwitterion to change as it moves through the pH gradient (such a buffer is called an *ampholyte*). For each protein, there is a pH at which the number of positively charged sites is equal to the number of negatively charged sites, giving a net charge of zero. At this point in the pH gradient, the electrophoretic mobility is zero as the protein has no net pull to move. This is called the *isoelectric point* (pI).

2.1.2. SDS-PAGE

A protein can be denatured by heating it in the presence of SDS, an ionized detergent. The heat causes the protein to unfold into a long strand, and allows the detergent to form a *micelle* (molecular cage) around it. This micelle behaves like a long rod with a surface charge proportional to its length, which is proportional to the protein's molecular weight and much larger than any net charge inherent in the macromolecule itself. As the charge is now proportional to the molecular weight, the amino-acid sequence loses its importance in determining migration, and mobility is then inversely proportional to the radius of the molecule. This is the basis for SDS-gel electrophoresis. By equating the volume of a sphere of radius r with the volume of a rod of length L and radius b, the spherical frictional coefficient of the rod can be derived (*6*), and the electrophoretic mobility of a rod-shaped molecule is

$$\frac{c\left[\log\left(\dfrac{L}{b}\right) - 0.30\right]}{3\pi\eta}$$

where *c* is a proportionality constant relating micellar charge to molecular length. This implies that very long molecules have similar mobilities because of the flatness of the logarithm function.

The electrophoresis buffer carries the current and determines the pH of the medium. A buffer is an ionic solution of a weak acid (or a weak base) plus its salt, and functions to maintain a constant pH. The relationship is described by the Henderson–Hasselbalch equation:

$$pH = pK_a + \log_{10}\left(\frac{\text{salt}}{\text{acid}}\right)$$

where pK_a is a constant that depends on the acid (or base). When pH is near pK_a, the curve is relatively flat over a considerable range of the ratio. A constant pH is important because the conformation of the macromolecules can be drastically affected by a change in pH: proteins return to their folded configuration. As a result, the molecule is no longer a rod and moves more slowly through the gel than would be expected based on its molecular weight alone. As various proteins return to their natural configuration at different pH values, a stable buffer is important for reliable SDS-PAGE.

The ionic strength of the buffer impacts mobility. At high ionic concentration, the ions hinder the movement of the protein by forming a cloud around the macromolecule. Different buffers (e.g., with varying concentration of Tris) are appropriate depending on the size of the molecule. If smaller molecules are to be separated, higher buffer concentrations are used to improve resolution, while with larger molecules the buffer concentration is lowered to reduce the time required to run the gel.

2.1.3. 2D-PAGE

Kenrick and Margolis *(9)* did the initial work on two-dimensional electrophoresis by combining isoelectric focusing with SDS-gel electrophoresis. Subsequently, Klose *(10)*, O'Farrel *(11)*, and Scheele *(12)* each published applications of this method which form the basis for current techniques of 2D-PAGE. Isoelectric focusing is applied in one direction to separate proteins based on their pI. The gel is then soaked in SDS, and the resultant bands are exposed to an electric field, perpendicular to the first one. With this technique, several thousand proteins can be isolated on a single gel. Considerable effort has been invested to improve the resolution and reproducibility of these gels to maximize the number of proteins detected and to reduce the within- and between-laboratory variability.

2.2. Gel Preparation

In polyacrylamide gels, the acrylamide monomer and a crosslinking agent such as bisacrylamide are mixed and react in the presence of other reagents to form an acrylamide polymer *(13)*. The rate at which molecules pass through a gel is determined by the pore size of the gel. For a given pore size, larger molecules will travel more slowly. The pore size of the gel is determined by the percentage of crosslinker it contains. Gels are characterized by the total percentage of acrylamides (linear and crosslinker) and the percentage of crosslinker. In addition to a separating gel, a stacking gel is used to concentrate the proteins into a stack of very thin bands before they reach the separating gel. The proteins are loaded onto a gel by placing them in a well that is several millimeters deep. If they are immediately placed onto the separating gel, proteins at the top of the well have significantly further to travel before they enter the gel than those at the bottom of the well. To account for this, the stacking gel concentrates all proteins into a very thin band (<1 mm thick). A spacer gel is sometimes used between the stacking and the separating gels when large concentrations of proteins are to be separated.

An early method for isoelectric focusing used carrier ampholytes *(8,14)*. These small molecules, polyaminocarboxylic acids, rapidly migrate to a location in the gel that corresponds to the isoelectric point of the ampholyte. The mixture of ampholytes with varying pI sets up a pH gradient before the proteins have migrated very far. The proteins then move to their respective isoelectric point. This method has several drawbacks. After stabilization, the ampholytes tend to drift toward the cathode, causing the gradient to vary. These gradients are also distorted by high salt or high protein concentrations. Furthermore, the gels become stretched when extruded from the glass tubes used to hold the gel. An alternative method, the immobilized pH gradient (IPG) method, was developed by Bjellqvist et al. *(15)*. For this method, the pH gradient is created by covalently binding molecules to the acrylamide gel, allowing this gradient to be tailored to the problem. The gradient is typically a linear, step, or sigmoidal function of location. These gels are now commercially available, thereby reducing much of the variability in the gradients.

SDS-PAGE can be run either vertically or horizontally. Neither method offers an advantage with respect to reproducibility and both require stacking gels. By convention, PAGE is usually performed vertically and agarose gel electrophoresis horizontally. With precast gels, this is purely for historical reasons. However, acrylamide does not polymerize completely in contact with air. Consequently, when these gels are poured in the laboratory (as opposed to utilizing precast commercial gels), air needs to be excluded from the surface, usually by placing a layer of water or mineral oil over the unset acrylamide. As it is convenient to have as small an amount of this liquid as possible, the gel is poured vertically and a layer of water is gently instilled on top. By contrast, agarose gels set by cooling at room temperature; the molten gel is poured into a horizontal mold so the entire surface can cool rapidly by heat radiating into the air. As commercial gels are usually poured in an inert atmosphere and are ordered set, it is unimportant whether they are run vertically or horizontally.

2.3. Sample Preparation

Protein samples can be obtained from any cell, tissue, or protein-containing fluid. Special preparations may be required to break open cells (*lysis*) or to extract proteins from membrane structures. Most preparations include a buffer (to control pH and provide electrolytes for the current), detergents and chaotropic agents (to denature the protein and separate monomeric subunits), reducing agents (to break disulfide bonds), and a tracking dye (so that the progress of the electrophoresis can be monitored). If the sample is contaminated with nucleic acids, enzymes called *endonucleases* may be necessary to digest them, as nucleic-acid contamination can alter the electrophoretic characteristics of certain proteins.

2.4. 2D Gel Electrophoresis Procedure

In the first dimension, the proteins are separated by charge. For example, a gel strip, a few millimeters thick and containing a nonlinear (sigmoidal) immobilized pH gradient (3.5–10.0), can be made in the laboratory or purchased commercially for this purpose (*16*). The gel is connected to an electric current via the appropriate electrophoresis apparatus. After the sample is placed at the cathode end, the electric field is applied. The voltage is linearly increased from 300 to 3500 V over 3 h, then left at 3500 V for 3 h, and then at 5000 V for up to 100 kVh. After the run, the disulfide bonds are reduced chemically to prevent them holding the protein in a folded structure.

A vertical slab gel is typically used for the second dimension (*17*). The IPG strip is trimmed at the ends and placed in a solution at the top of the gel where they fuse. The gel is run at 40 mA with 100–400 V at 8–12°C for 5 h. A horizontal apparatus may be less efficient in transferring the protein from the first gel to the second gel, especially if the fusion is incomplete. Low-concentration proteins can be lost in the transfer.

2.5. Gel Staining and Scanning

Radiolabeling is the most sensitive method for localizing proteins on gels (*14*). This involves the binding of radioactive isotopes to the macromolecules and detecting, usually with film or phosphorescent screen, the radiation produced. With this method proteins can be detected at 20 parts per million. Chemoilluminescence and chemoilluminescence silver staining are alternative nonradioactive methods that are 100 times more sensitive than an organic dye such as Coomassie brilliant blue. Other stains include amido black, Ponceau S, and bromophenol blue. The silver staining method is started immediately after the second run. The gel is washed, bathed in a soaking solution for several hours, and chemically treated. Then the staining solution is applied for 30 min, washed, and developed. Fluorescent stains can also be used and require a much shorter processing time because the fluorescence peaks after a relatively short time.

A laser densitometer is used to detect the concentration of staining. Each stain has a specific wavelength at which it emits the maximum signal. Improperly adjusted densitometers may produce peaks that are off the scale or eliminate short peaks when the background is improperly subtracted. This creates right and left censoring, respectively. The integral of each peak gives a quantity proportional to the abundance of the protein. Unfortunately, because the relationship between the intensity of staining and the protein concentration depends on the interaction between the protein and staining agent, in general, absolute concentrations cannot be calculated.

2.6. Spot Identification

Proteins can be identified by a number of methods *(8,14)*. These include the determination of amino-acid composition and peptide mass fingerprinting. With several methods the individual proteins need to be transferred from the gel to another medium. Electroblotting to polyvinyldifluoride (PVDF) membranes is frequently used. Blotting can also be performed by semidry electrophoresis, vacuum, or capillary action. The advantage of semidry blotting is that it takes less than half an hour compared to other transfers that take several hours. The PVDF membranes are stained to locate the proteins, which are removed with a razor blade for identification. A minimum of 250 ng of protein per spot is required for identification. When analysis is performed using high-performance liquid chromatography (HPLC), the distribution of the amino acids, the pI, and the molecular weight are used to identify the protein from databases using a least-squares distance metric. An automated sequencer can only identify one protein per day while the HPLC method is 5–10 times faster. For peptide mass fingerprinting, the proteins are digested in the PVDF membrane using enzymes and analyzed using mass spectroscopy. The mass spectrum is then matched with a library of peptides. The ultimate goal for protein identification is to use standard maps overlaid on 2D-PAGE images, but these images require resource-intensive methods and sophisticated computer algorithms to develop the maps.

2.7. Other Sources of Variation

Protein separation and identification require many technical steps such as those described previously. Every step leads to (possibly high) variation in the output if not properly performed. A few additional sources of variation are described here:

1. Buffers are good growth media for microorganisms, which contain abundant protein. These buffers need to be stored in tight containers and refrigerated to inhibit growth. Periodically, they need to be replaced.
2. Care must be taken in sample collection, handling, and storage to avoid sample contamination. For instance, cells must be removed immediately from fluid samples. Proteins should not be left at room temperature and can be stored at 2–8°C for up to 3 d but at –20°C for longer periods. Thawing and refreezing should be minimized to avoid damaging the proteins. Vibration and pressure changes during transportation can also damage the proteins.
3. Consistency in the experimental protocol is of paramount importance. For instance, if too little sample is used, the small peaks will be below the limit of detection. Likewise, if too much sample is used, the peaks will be blunted, creating the same censoring problem as densitometer maladjustment.

3. State-of-the-Art Analytical Methods

We now review the analytical methods implemented in software packages such as MELANIE *(18–20)* and HERMeS *(21–25)*. Let $I(x,y)$ denote the two-dimensional image, and, by convention, the larger $I(x,y)$ is, the darker the pixel is.

3.1. Filtering Gel Images

To reduce the high-frequency background noise, the signal is extracted by applying a smoothing filter. The most popular filters are

1. *Gaussian smoothing*, which convolves the image with the operator

$$\frac{1}{16}\begin{pmatrix} 1 & 2 & 1 \\ 2 & 4 & 2 \\ 1 & 2 & 1 \end{pmatrix},$$

2. *diffusion smoothing*, that is,

$$I^{(t+1)}(x,y) = \frac{1}{2}\left(I^{(t+1)}(x-1,y) + I^{(t)}(x+1,y)\right)$$

$$I^{(t+2)}(x,y) = \frac{1}{2}\left(I^{(t+1)}(x-1,y) + I^{(t+2)}(x+1,y)\right)$$

$$I^{(t+3)}(x,y) = \frac{1}{2}\left(I^{(t+3)}(x,y-1) + I^{(t+2)}(x,y+1)\right)$$

$$I^{(t+4)}(x,y) = \frac{1}{2}\left(I^{(t+3)}(x,y-1) + I^{(t+4)}(x,y+1)\right),$$

3. *polynomial smoothing*, where the pixel intensities in a small area (e.g., 3×3, 7×7) are approximated by a second-degree polynomial function in x and y

4. *adaptive smoothing*, that is,

$$I^{(t+1)}(x,y) = \frac{1}{N^{(t)}} \sum_{i=-1}^{1} \sum_{j=-1}^{1} I^{(t)}(x+i,y+j)\, w^{(t)}(x+i,y+j)$$

with

$$w^{(t)}(x,y) = \exp\left(-\frac{\left(d^{(t)}(x,y)\right)^2}{2\,K^2}\right),\ N^{(t)} = \sum_{i=-1}^{1}\sum_{j=-1}^{1} w^{(t)}(x+i,y+j),\ d^{(t)}(x,y) = \sqrt{G_x^2 + G_y^2}$$

where G_x and G_y are the gradients along the *x*- and *y*-axis, respectively *(20)*.

Gaussian deconvolution is an alternative approach to remove noise and blur *(2)*. Each spot is modeled as

$$I(x,y) = \sum_{k=-m}^{m} A(x+k,y)\, g_k + e(x,y) \text{ with } A(x+k,y) \geq 0 \text{ and } g_k = \frac{1}{\sqrt{2\pi\sigma^2}} \exp\left(\frac{-k^2}{2\sigma^2}\right)$$

where *m* is the integer part of 3σ and $e(x, y)$ represents random noise. Estimates are obtained via the constrained least-squares procedure. This approach tends to be overly sensitive to noise and to oversplit spots *(26)*.

3.2. Spot Detection

For automatic spot detection, nonparametric procedures, based either on second derivatives *(19,20)* or on mathematical morphology *(2)*, are utilized.

Let $p = (x, y)$ be a point on the image, S_i a spot, and T the saturation threshold

$$\max(I) - \frac{100 - saturation}{100}\left(\max(I) - \min(I)\right)$$

where $0 \leq saturation \leq 100$ (*saturation* = 100 when no pixel is saturated), and $\Delta I(p)$ the Laplacian

$$-\left(\frac{\partial^2}{\partial x^2} I(p) + \frac{\partial^2}{\partial y^2} I(p)\right) .$$

Is *p* part of the spot S_i? Select thresholds *l*, *r*, *c* (i.e., small positive constants). If $I(p) < T$,

$$p \in S_i \quad \Leftrightarrow \quad \min\left(\frac{\partial^2}{\partial x^2} I(p) - r, \frac{\partial^2}{\partial y^2} I(p) - c\right) > 0$$

when $-\Delta I(p) - l \geq 0$. If $I(p) > T$,

$$p \in S_i \quad \Leftrightarrow \quad \min\left(\frac{\partial^2}{\partial x^2} I(p)\ ,\ \frac{\partial^2}{\partial y^2} I(p)\right) > 0 .$$

Small values of *l* allow the detection of as many spots as possible, while high values only yield dark spots with flat spots being ignored. High values of *r* and *c* help separate spots that are in close proximity of each other and to eliminate streaks. The algorithm identifies the spots by searching for the most negative values of the Laplacian and the two second derivatives *(19,20)*. The Laplacian is indeed most negative at local peaks, while at the inflection points between unresolved spots the minimum value of the two second derivatives is negative.

With mathematical morphology, one can study characteristics of objects by investigating whether a standard shape fits into them. In this context, one searches for elevations relative to the local background brightness, and constructs an image based on the heights of these elevations. This is achieved by subtracting the *closing* of the image from the original image. This is the so-called *top-hat transform*, which in essence assesses whether a cylinder of a chosen radius fits into the elevations. To obtain the closing of the image, one first replaces each pixel value $I(x, y)$ with the local minimum intensity in a disk around each pixel, and then one replaces the resulting pixel value $I'(x, y)$ with a local maximum intensity, that is,

$$I''(x, y) = \max_{k,l} I'(x+k, y+l)$$

where $\sqrt{k^2 + l^2} \leq R$ (the disc radius) and $I'(x, y) = \min_{k,l} I(x+k, y+l)$.

In this closing, only pixels within elevations narrower than the chosen disc size have changed their values from the original image, and thus will show when the two images are subtracted. The radius R is selected to be the smallest value so that the disk is larger than the smallest spot *(2)*. Shapes (or *structuring elements*) other than disks can be considered (e.g., spheres *[27]*).

Alternatively, instead of a fixed structuring element, one can look for all *h*-domes *(21,28,29)*. An *h*-dome is a connected region of pixels with intensity above *h* and greater than any pixel bordering the *h*-dome. These *h*-domes are not constrained by size. Algorithms for searching for these regions are more complex than the top-hat transform. The choice of *h* is crucial: if it is too small, background streaks will be recovered rather than spots, and, if it is too large, narrow peaks due to high-amplitude noise will be selected. Raising *h* stepwise allows the resolution of overlapping spots *(28)*.

3.3. Background Filtering

The smooth background noise, which consists of vertical and horizontal streaks, is removed, either by subtracting the global minimum pixel value from all pixels or by estimating the background outside the spots with a third-order polynomial function *(20)*. Because the background varies significantly across the image, a single threshold tends to work poorly: when it is set too high, faint spots are lost and, when it is set too low, high background is regarded as signal *(28)*.

Mathematical morphology has also been utilized to remove the streaks *(21,27)*. One subtracts from the original image its closing with respect to two structuring elements, one vertical and one horizontal bar of one-pixel width, the lengths of which are slightly greater than those of the vertical and horizontal extents of the largest spots.

3.4. Spot Quantification

Spot characteristics are estimated by fitting two-dimensional Gaussian curves via the least-squares method:

$$g(x,y) = A \, \exp\left\{ \left(\frac{x - x_c}{\sigma_X} \right)^2 + \left(\frac{y - y_c}{\sigma_Y} \right)^2 \right\} + B$$

with A representing the amplitude, (x_c, y_c) the center, the σ's the spread along the principal axes and B the background level *(20,30,31)*. Spot models based on two half-Gaussian curves have also been utilized *(21)*. However, many spots are not Gaussian in nature because of several factors: local inhomogeneity within the acrylamide, overloading of the sample within the gel, adsorption of some proteins onto the acrylamide matrix, failure of some polypeptides to focus in the first dimension, and tendency for chemically distinct but barely resolved proteins to displace each other *(26)*.

Spot characteristics are thus better estimated directly:

$$\text{area} = \text{AREA} = \text{number of pixels} \times \text{pixel area}$$

$$\text{optical density} = \text{OD} = \max_{x,y \in \text{spot}} I(x,y)$$

$$\text{percent optical density} = \%\text{OD} = 100 \times \frac{\text{OD}}{\sum_{s=1}^{n} \text{OD}_s}$$

$$\text{volume} = \text{VOL} = \sum_{x,y \in \text{spot}} I(x,y)$$

$$\text{percent volume} = \%\text{VOL} = 100 \times \frac{\text{VOL}}{\sum_{s=1}^{n} \text{VOL}_s}$$

where OD_s and VOL_s are, respectively, the optical density and the volume of spot s in a gel containing n spots.

3.5. Image Alignment

Gels are aligned via polynomial image warping. Identify landmarks (or control points) on each image and choose one gel as the reference gel. The alignment algorithm attempts to superimpose these landmarks by stretching and shrinking the images. Let (x,y) be the pixel coordinates in the original image and $[u(x),$ $v(y)]$ those in the warped image. The latter are first-, second- or third-order univariable polynomials or their inverses. Estimate the parameters of these polynomial functions via the least-squares criterion by summing over the landmarks. Specifically, let there be M landmarks on each gel. The parameters are obtained by minimizing, for instance, if $M \geq 4$,

$$\sum_{i=1}^{M} \left(u_i - (a_0 + a_1 x_i + a_2 x_i^2 + a_3 x_i^3)\right)^2 \quad \text{and} \quad \sum_{i=1}^{M} \left(v_i - (b_0 + b_1 y_i + b_2 y_i^2 + b_3 y_i^3)\right)^2$$

where (u_i, v_i) refers to landmark i on the reference gel and (x_i, y_i) to its position on another gel. The value M determines the order of the polynomial: a polynomial of degree n requires at least $n + 1$ landmarks.

3.6. Spot Matching

Local gel-to-gel variations make it impossible to utilize a single transformation to map the spots from one gel to another. One approach is to divide the image into a number of small rectangular regions and to select, in each segment, 3 or more evenly spaced spots as reference points (28). These reference points serve to compute a transformation that maps spot centers from one film to another. Spots are considered matched if the transformed spot center from one gel and the corresponding spot center on the other gel are within 0.8 mm [a slightly more stringent criterion of 0.7 mm has also been utilized (32)]. This procedure works best for the area defined by the reference points: spots located at the edges of the rectangular regions can be poorly matched. As a remedy, the following steps are added to the procedure: triangles of nearby matched spots are considered on both images, and the above algorithm is applied to the yet unmatched spots within these triangles.

Alternatively, for each spot on a gel, consider a cluster of neighboring spots (20). The central spot is regarded as the primary spot, and the surrounding spots as secondary spots. A spot belongs to a cluster if its centroid is inside a circle of fixed radius. This radius depends on the image dimension, number of spots on a gel, and minimum number of spots in the cluster. The clusters are characterized by polar coordinates centered at the primary spot. First, match the clusters with highest-intensity primary spots. Compare clusters via a probabilistic similarity measure. The probability that the next random hit falls within a cluster where $m - 1$ spots have been matched is given by

$$p_m = \frac{A_s - A_{m-1}}{A_c - A_{m-1}}$$

where A_s is the sum of the secondary areas in the cluster, A_c the total area within the boundary of the cluster, and A_{m-1} the total area of matched spots. If N stands for the number of spots in one cluster,

Prob(at least m spots are matched in N trials)

$$= \sum_{h=m}^{N} \binom{N}{h} \prod_{i=1}^{h} p_i \prod_{i=h+1}^{N} (1-p_i) \approx \sum_{h=m}^{N} \binom{N}{h} p_G^h (1-p_G)^{N-h} \quad \text{where } p_G = \left(\prod_{i=1}^{m} p_i \right)^{1/m}.$$

That spots be reliably matched is of paramount importance in the creation of representative images and in subsequent pattern-recognition analyses. Proceed with a consistency check for possible mismatching:

$$\binom{u}{v} = \binom{t_x}{t_y} + \begin{pmatrix} A & B \\ C & D \end{pmatrix} \binom{x}{y}$$

by ensuring that $L = AD - BC \approx 1$ (rotation). For each primary cluster, estimate parameters from each set of 3 matched spots. When $L = 1.0 \pm 0.25$ and the rotation angle is ± 10 degrees, the pairing is declared suitable. To project the remaining spots in the clusters, estimate the rotation parameters $A, \dots, D$ by the least-squares method from good matchings in the two clusters.

Artificial-intelligence methodology has also been used to match spot lists *(23)*. Because they are not based on geometrical considerations, they should be able to cope better with discontinuous gel distortions. Spot clusters are described via the angles and distances between any two spots in the cluster. Distances are divided into 3 classes and angles into 16 classes. Measurements are coded via their class identifiers. Spots are then matched via syntactic pattern-recognition techniques. Heuristic rules are imposed to limit the number of searches. Isolated spots tend to be problematic for this approach.

3.7. Creating Synthetic Gels

To obtain a master image from at least 3 pairwise matched gels, first select a reference gel *(20)*. Check that the spots on the reference gel are well matched to spots on two other gels. These form triangles of matched spots, that is, the *starting groups*. Extend the starting groups by adding spots using the connectivity test: a spot must be matched with at least one other spot in the initial group. When all spots on the reference gel have been considered, create additional groups with the spots on the second gel that are not part of a group. Repeat with the other gels.

The synthetic gel contains the same number of spots as there are determined groups (these are the representative spots). The position of a spot on the synthetic gel is taken from the reference gel if the group has a spot on the reference gel. Otherwise, one translates the coordinates of the closest spot by considering a set of neighbors that have representatives on the reference gel. The intensity of the master spot is the average over the spots in the group. Its shape is the shape of the spot in the group that is closest in area to the average of the group *(20)*.

To ensure the reliability of the master gel, some investigators compute an overlap measure between each master spot and the corresponding spots (considered one at a time) on the aligned gels *(1)*. This in effect assesses the quality of the matches on which the master gel relies. The overlap measure is simply the value of a Gaussian function evaluated at the physical distance of the spot centers. This Gaussian function is chosen to have height of 1.0 and width depending on the matching criterion: for instance, 0.7 mm if spots on different aligned gels are considered matched when they are within 0.7 mm *(32)*.

3.8. Pattern Recognition

For pattern-recognition purposes, only spots that yield highly reliable features on the master image are considered in the analyses. For instance, the overlap measure between the master spot and a corresponding object on one of the aligned gels (cf. **Subheading 3.7.**) needs to be above a specific threshold (typically 0.5) and to exceed 90% of the largest overlap of the master spot with the original spots or of the overlap of the gel spot with all the master spots *(1)*.

To find significant protein patterns associated with a specific disease, investigators have first recourse to principal-component analysis *(1,25)* or correspondence analysis and factor analysis to reduce dimensionality *(18–20)*. This requires the computation of the normalized observation table so that the columns have mean 0 and variance 1. In factor analysis, the eigenvalues and eigenvectors of the covariance matrix are extracted to determine a factorial space (usually of dimension between 1 and 3). The gels are projected as points onto the factorial space. Spots can also be mapped onto this space so that characteristic spots can be identified: spots fall within the cluster of gels they typify.

These authors then apply clustering algorithms to the transformed data. The difficulty here is to define a meaningful distance metric. With principal-component analysis, one candidate is the Euclidean distance in the transformed space after weighing each coordinate by the percentage of the total variance represented by the corresponding principal component *(1)*. The usual hierarchical clustering procedures, based on complete or single or average linkage, are utilized *(1,18–20,33)*. A heuristic clustering algorithm has also been proposed *(18,19)*. Suppose that *n* gels are to be classified into *k* classes. Select *k* gels by maximizing the Euclidean distance between them. These define *k* classes. A heuristic search is then performed: each of the remaining *n – k* gels is included into one class and class descriptions are formulated. Iterate this process by choosing one gel per class, excluding the first *k* gels, to form new classes and repeating the previous step. This process continues until the classification converges.

One is in effect searching for protein patterns that best distinguish two groups of images (one from a "disease" group and one from a "control" group). Classification procedures would be best suited for this purpose. In actuality, patterns are identified, ignoring the associated outcomes, and then the inferred patterns are reconciled to the known groups.

4. Brief Introduction to Wavelets

Wavelets are building-block functions like *sine* and *cosine* functions in the Fourier transform *(34)*. They oscillate about 0 and dampen to 0. This localization in time or space renders them highly versatile to model signals with nonsmooth features or that vary over time or space. The *father wavelet* or *scaling function* ϕ represents smooth, low-frequency components while the *mother wavelet* ψ represents detail, high-frequency components:

$$\int_{-\infty}^{+\infty}\phi(t)=1 \quad \text{and} \quad \int_{-\infty}^{+\infty}\varphi(t)=0 \ .$$

A number of orthogonal wavelet families have been constructed: for instance, Haar wavelets (symmetric square waves with compact support), daublets (continuous waves with compact support, d2 to d20 in S-plus *[35]*), symmlets (nearly symmetric waves with compact support, s4 to s20 in S-plus *[35]*).

Through a multiresolution analysis, one obtains fine to coarse resolution (scale) components of the signal, that is, for a one-dimensional signal,

$$f(t)=\sum_k s_{J,k}\phi_{J,k}(t)+\sum_k d_{J,k}\varphi_{J,k}(t)+\sum_k d_{J-1,k}\varphi_{J-1,k}(t)+\ldots+\sum_k d_{1,k}\varphi_{1,k}(t)$$

where J is the number of multiresolution components considered. The functions $\phi_{j,k}(t)$ and $\psi_{j,k}(t)$ are generated from ϕ and ψ by scaling and translation, that is,

$$\phi_{j,k}(t)=2^{\frac{-j}{2}}\phi(2^{-j}t-k)=2^{\frac{-j}{2}}\phi\left(\frac{t-2^j k}{2^j}\right) \quad \text{and} \quad \varphi_{j,k}(t)=2^{\frac{-j}{2}}\varphi(2^{-j}t-k)=2^{\frac{-j}{2}}\varphi\left(\frac{t-2^j k}{2^j}\right).$$

The scale/dilation factor 2^j affects the width of $\phi_{j,k}(t)$ and $\psi_{j,k}(t)$. The translation/location parameter $2^j k$ is coupled to the scale factor: as the support of $\phi_{j,k}(t)$ and $\psi_{j,k}(t)$ gets wider the translation steps become larger. As 2^j increases, $\phi_{j,k}(t)$ and $\psi_{j,k}(t)$ become shorter and more spread out. Finally, $s_{J,k}$, $d_{J,k}$, ..., $d_{1,k}$ are the wavelet-transform coefficients:

$$\text{scaling function coefficients} \qquad s_{J,k} \ = \ \int_{-\infty}^{+\infty}f(t)\,\phi_{J,k}(t)\ dt$$

$$\text{wavelet coefficients} \qquad d_{j,k} \ = \ \int_{-\infty}^{+\infty}f(t)\,\varphi_{j,k}(t)\ dt$$

The $\phi_{j,k}(t)$'s and $\psi_{j,k}(t)$'s form an orthogonal basis:

$$\int_{-\infty}^{+\infty}\phi_{J,k}(t)\,\phi_{J,k'}(t)\,dt=\delta_{k,k'}\ , \quad \int_{-\infty}^{+\infty}\varphi_{J,k}(t)\,\phi_{J,k'}(t)\,dt=0\ , \quad \int_{-\infty}^{+\infty}\varphi_{j,k}(t)\,\varphi_{j,k'}(t)\,dt=\delta_{k,k'}\delta_j,$$

where $\delta_{i,j}=1$ if $i=j$ and 0 if $i\neq j$.

The discrete wavelet transform $\mathbf{W}$ for the discrete signal $\mathbf{f}=(f_1,f_2,\ldots,f_n)'$ is defined as

$$\mathbf{w}=\mathbf{W}\,\mathbf{f} \quad \text{where} \quad \mathbf{w}'=(\ s_J'\ d_J'\ d_{J-1}'\ \ldots\ d_1')$$

with

$$\mathbf{s_J} = (s_{J,1}, s_{J,2}, \ldots, s_{J,n/2^J})', \mathbf{d_J} = (d_{J,1}, d_{J,2}, \ldots, d_{J,n/2^J})', \mathbf{d_{J-1}} = (d_{J-1,1}, d_{J-1,2}, \ldots, d_{J-1,n/2^{J-1}})', \ldots,$$
$$\mathbf{d_1} = (d_{1,1}, d_{1,2}, \ldots, d_{1,n/2})'.$$

Each of the so-called crystals $\mathbf{s_J}$, $\mathbf{d_J}$, $\mathbf{d_{J-1}}$, ... , $\mathbf{d_1}$ contains the coefficients corresponding to a set of translated wavelet functions. In the multiresolution analysis,

$$f(t) \approx S_J(t) + D_J(t) + D_{J-1}(t) + \ldots + D_1(t),$$

the smooth and detail signals are represented, respectively, by

$$S_J(t) = \sum_k s_{J,k} \phi_{J,k}(t) \quad \text{and} \quad D_j(t) = \sum_k s_{j,k} \varphi_{j,k}(t) \quad j = 1, \ldots, J$$

To compress an image, one utilizes a two-dimensional wavelet family

$$\Phi(x,y) = \phi_h(x) \times \phi_v(y) = \text{horizontal father} \times \text{vertical father}$$

$$\Psi^v(x,y) = \psi_h(x) \times \phi_v(y) = \text{horizontal mother} \times \text{vertical father}$$

$$\Psi^h(x,y) = \phi_h(x) \times \psi_v(y) = \text{horizontal father} \times \text{vertical mother}$$

$$\Psi^d(x,y) = \psi_h(x) \times \psi_v(y) = \text{horizontal mother} \times \text{vertical mother}.$$

The father wavelet Φ deals with the smooth aspect and the mother wavelets Y deal with the details in the vertical (Ψ^v), horizontal (Ψ^h), and diagonal (Ψ^d) dimensions. (**Figure 2** shows the diagonal s8 wavelet.) The two-dimensional wavelet approximation is then

$$F(x,y) \approx \sum_{m,n} s_{J,m,n} \Phi_{J,m,n}(x,y) + \sum_{j=1}^{J} \sum_{m,n} v_{j,m,n} \Psi^v_{j,m,n}(x,y) + \sum_{j=1}^{J} \sum_{m,n} h_{j,m,n} \Psi^h_{j,m,n}(x,y)$$

$$+ \sum_{j=1}^{J} \sum_{m,n} d_{j,m,n} \Psi^d_{j,m,n}(x,y)$$

with

$$\Phi_{J,m,n}(x,y) = 2^{-J} \Phi(2^{-J}x - m, 2^{-J}y - n), \Psi^v_{j,m,n}(x,y) = 2^{-j} \Psi^v(2^{-j}x - m, 2^{-j}y - n),$$

$$\Psi^h_{j,m,n}(x,y) = \Psi^h(2^{-j}x - m, 2^{-j}y - n), \Psi^d_{j,m,n}(x,y) = \Psi^d(2^{-j}x - m, 2^{-j}y - n),$$

$$s_{J,m,n} = \int\int \Phi_{J,m,n}(x,y) F(x,y) \, dx \, dy, \quad v_{j,m,n} = \int\int \Psi^v_{j,m,n}(x,y) F(x,y) \, dx \, dy,$$

$$h_{j,m,n} = \int\int \Psi^h_{j,m,n}(x,y) F(x,y) \, dx \, dy, \quad d_{j,m,n} = \int\int \Psi^d_{j,m,n}(x,y) F(x,y) \, dx \, dy.$$

The two-dimensional discrete wavelet transform maps an $m \times n$ discrete image to $m \times n$ matrix of wavelet coefficients $\mathbf{w}_{m,n}$. In S-plus *(35)*, $\mathbf{w}_{m,n}$ is decomposed into submatrices with coefficients for different multiresolution levels:

sJ – sJ   with coefficients $s_{J,m,n}$ for the smooth part
d1 – s1,..., dJ – sJ   with coefficients $v_{j,m,n}$ for the vertical detail

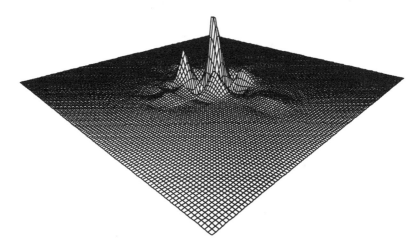

Fig. 2. s8 diagonal mother wavelet.

s1 − d1,..., sJ − dJ   with coefficients $h_{j,m,n}$ for the horizontal detail
d1 − d1,..., dJ − dJ   with coefficients $d_{j,m,n}$ for the diagonal detail.

Hence, in the multiresolution analysis,

$$F(x,y) \approx S_J(x,y) + \sum_{j=1}^{J} D_j^v(x,y) + \sum_{j=1}^{J} D_j^h(x,y) + \sum_{j=1}^{J} D_j^d(x,y),$$

where

$$S_J(x,y) = \sum_{m,n} s_{J,m,n}\, \Phi_{m,n}(x,y) \quad , \quad D_j^h(x,y) = \sum_{m,n} h_{j,m,n}\, \Psi_{m,n}^h(x,y) ,$$

$$D_j^v(x,y) = \sum_{m,n} v_{j,m,n}\, \Psi_{m,n}^h(x,y) \quad , \quad D_j^d(x,y) = \sum_{m,n} d_{j,m,n}\, \Psi_{m,n}^d(x,y) .$$

5. Wavelets for Two-Dimensional Electrophoretic Data

Few statistical techniques can cope with the high dimensionality of 2D-PAGE data. Hence, for analytical reasons, the information is often reduced to the volumes of a manageable set of selected spots. This prohibits exploratory investigations of the data for the purpose of formulating testable hypotheses. We explored the possibility of fitting Gaussian curves. We selected a number of spots from the gel depicted in **Fig. 1** and assessed via statistical tests that we would not be justified in assuming that their shape is Gaussian (as is clearly evidenced by **Fig. 3**). We turned to wavelets for their versatility in representing irregular signals.

With this methodology, much effort is needed to identify the most suitable representation. Once the coefficients are selected, mainstream techniques can

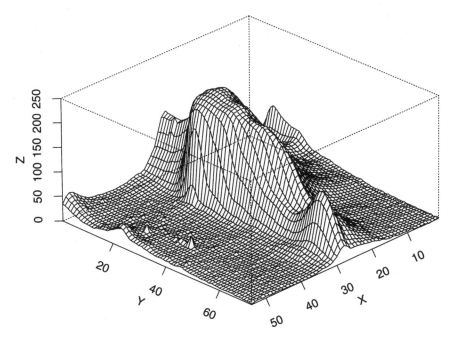

Fig. 3. Detail from kidney gel.

be applied to investigate the scientific questions of interest. We now review a few of these issues.

5.1. What Is the Most Suitable Wavelet Family to Represent Gels?

We considered Haar wavelets, daublets, and symmlets. All seemed suitable with the Haar family giving slightly worse results (**Figs. 4 – 7**). Even with an image corrupted by noise (Gaussian or Poisson), the reconstruction was highly successful (**Figs. 8 – 10**).

5.2. What Multiresolution Level Should One Select?

A high multiresolution level is not neccessary: most of the information is contained in the smoother crystals. The percentage of coefficients one wishes to retain is the more pertinent issue and depends on one's threshold for the volume of significant spots (**Figs. 11 – 14**).

5.3. How Does One Remove the Noise?

In the WaveShrink algorithm of S-plus (*35*), one applies the wavelet transform with J levels then shrinks the detail coefficients

$$\tilde{\mathbf{d}}_1 = \delta_{\lambda_1 \sigma_1}(\mathbf{d}_1), ..., \tilde{\mathbf{d}}_J = \delta_{\lambda_J \sigma_J}(\mathbf{d}_J)$$

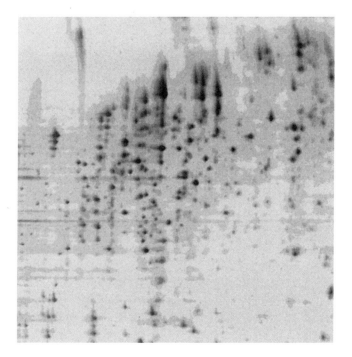

Fig. 4. Cropped image of kidney gel (512 × 512).

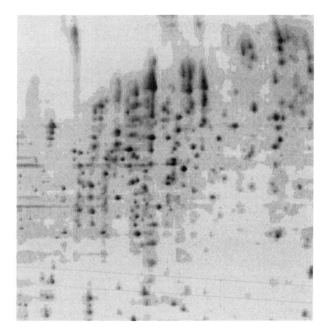

Fig. 5. Reconstruction of the cropped image of a kidney gel with daublets d4 at multiresolution level 4 and the largest 5% of the coefficients.

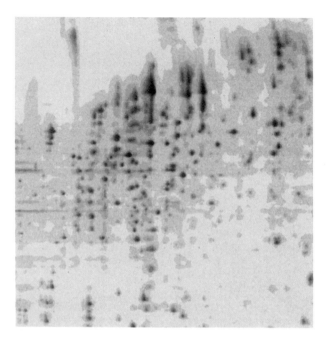

Fig. 6. Reconstruction of the cropped image of a kidney gel with symmlets s8 at multiresolution level 4 and the largest 5% of the coefficients.

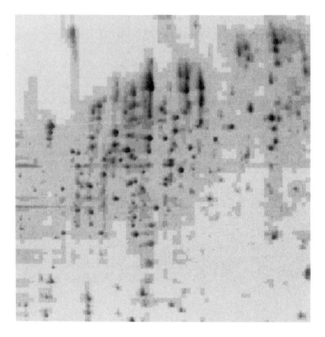

Fig. 7. Reconstruction of the cropped image of a kidney gel with Haar wavelets at multiresolution level 4 and the largest 5% of the coefficients.

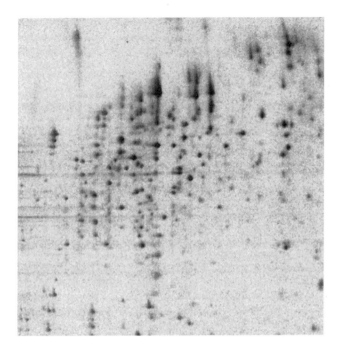

Fig. 8. Cropped gel image with added Gaussian noise (with variance 100).

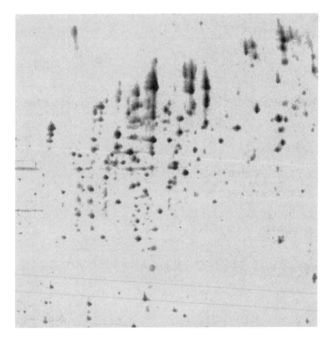

Fig. 9. Reconstruction of noisy gel with symmlets s8 at multiresolution level 1 and the largest 1% of the coefficients.

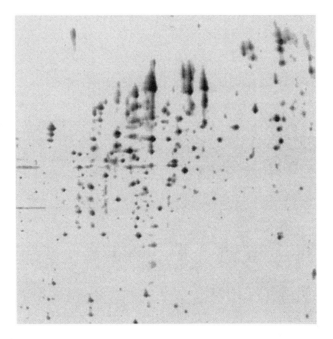

Fig. 10. Reconstruction of cropped image of kidney gel with symmlets s8 at multiresolution level 1 and the largest 5% of the coefficients.

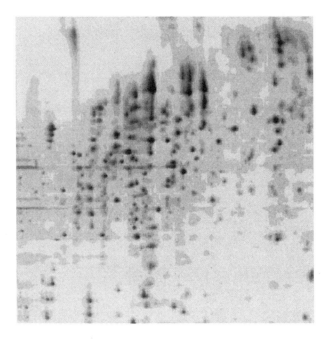

Fig. 11. Reconstruction of cropped image of kidney gel with daublets d4 at multiresolution level 3 and the largest 5% of the coefficients.

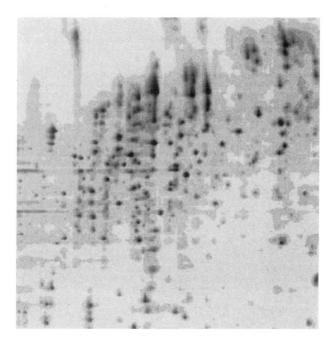

Fig. 12. Reconstruction of cropped image of kidney gel with daublets d4 at multiresolution level 5 and the largest 5% of the coefficients.

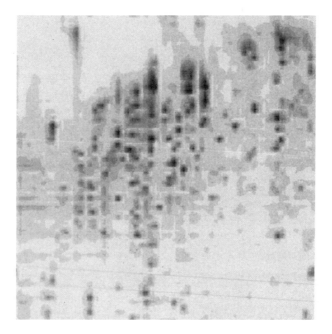

Fig. 13. Reconstruction of cropped image of kidney gel with daublets d4 at multiresolution level 3 and the largest 1% of the coefficients.

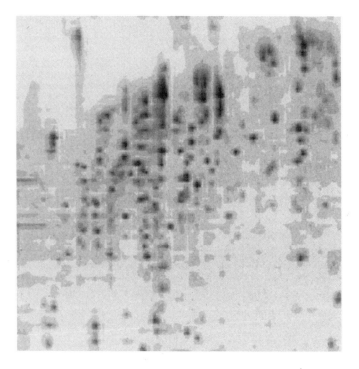

Fig. 14. Reconstruction of cropped image of kidney gel with daublets d4 at multiresolution level 5 and the largest 1% of the coefficients.

and reconstructs the image using $\tilde{\mathbf{d}}_1,...,\tilde{\mathbf{d}}_J,\mathbf{s}_J$. Shrinkage is performed using the so-called soft or hard shrinkage functions

$$\delta_\gamma^S(x) = \begin{cases} 0 & \text{if } |x| \le \gamma \\ \text{sign}(x)(|x|-\gamma) & \text{if } |x| > \gamma \end{cases} \quad , \quad \delta_\gamma^H(x) = \begin{cases} 0 & \text{if } |x| \le \gamma \\ x & \text{if } |x| > \gamma. \end{cases}$$

For the thresholds λ_j, one can select the so-called universal value

$$\lambda_j = \sqrt{2 \, \log n}$$

where n is the sample size. Alternatively, the value which minimizes the upper bound of the asymptotic risk (minimax) will result in less smoothing as it is always smaller than the universal threshold *(36)*. Finally, we also considered Stein's unbiased risk estimator (SURE) which is adapted to each multiresolution level; the threshold for $\mathbf{d_j}$ with K coefficients is

$$\lambda_j = \arg\min_{t \ge 0} \text{SURE}(\mathbf{d_j},t) \text{ where } \text{SURE}(\mathbf{d_j},t) = K - 2 \sum_{k=1}^{K} 1_{[d_{j,k} \le t\sigma_j]} + \sum_{k=1}^{K} \min\left((d_{j,k}\,\sigma_j)^2, t^2\right)$$

(37,38). To estimate the scale of the noise, that is, σ_j , one can rely either on the crystal corresponding to the finest detail or on all the detail crystals, or one can consider each crystal in turn, that is,

$$\tilde{\sigma}_j(\mathbf{d}_1), \ldots, \tilde{\sigma}_j(\mathbf{d}_1, \ldots, \mathbf{d}_J), \tilde{\sigma}_j(\mathbf{d}_J)$$

These algorithms yield rather poor results with the 2D-PAGE data (**Figs. 15–17**). Indeed, they are really smoothing techniques that are suitable to reduce highly localized and peaked noise. Here, the noise takes the form of streaks. All combinations we have tried result in the removal of important features.

We devised a hybrid procedure that seems to work well: hardshrinkage is utilized on the level 1 coefficients (these all tend to be very small) and the sJ - sJ crystal is multiplied by a constant between 0 and 1 (**Fig. 18**). This constant depends on the level of detail to be retained.

This routine is currently being optimized with respect to a biologically relevant objective criterion that involves the size of the spots being ignored.

5.4. How Does One Create a Master Gel?

Assume that the gels have been aligned. Wavelet coefficients are obtained for each of them. The synthetic gel is constructed by averaging the coefficients or by taking their median values. Variability in this construct can easily be computed.

5.5. How Does One Find Specific Protein Patterns for a Disease?

An analysis of the wavelet coefficients, for gels from diseased and control samples, based on classification and regression trees (CART), will highlight relevant clusters that best discriminate between the two groups. This has the advantage of considering both the location and the intensity of the spots simultaneously.

6. Conclusion

Electrophoresis has developed over the past 60 yr from a crude method able only to distinguish between very specific one-dimensional changes in experimental protocols to a highly complex technique. It is now possible not only to separate the genomic fingerprint of samples but also their proteome. While the technology has developed at an ever-increasing rate, the statistical techniques necessary to analyze such complex data structures has been left wanting. We have outlined some of the new methodologies that are currently available to take full advantage of the technology that is now in common usage in molecular biology laboratories.

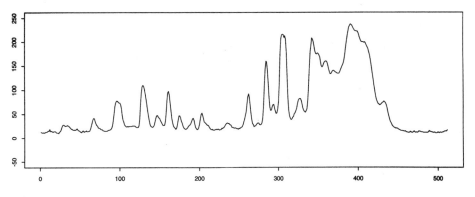

Fig. 15. Vertical slice of cropped image of kidney gel.

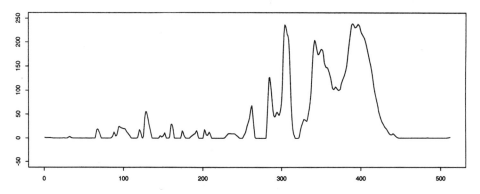

Fig. 16. Slice after hard shrinkage with universal threshold, estimating the scale of the noise separately for each crystal (multiresolution level 4).

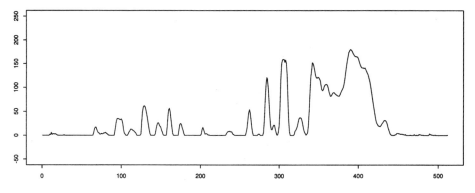

Fig. 17. Slice after soft shrinkage of the sJ - sJ crystal with universal threshold, estimating the scale of the noise from the sJ - sJ crystal (multiresolution level 4).

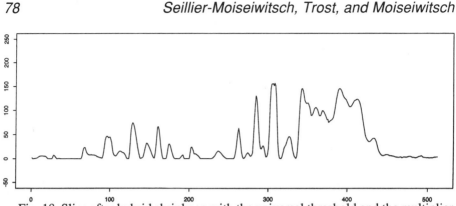

Fig. 18. Slice after hybrid shrinkage with the universal threshold and the multiplier set to 0.25 (multiresolution level 4).

References

1. Anderson, N. L., Hofmann, J. P., Gemmell, A., and Taylor, J. (1984) Global approaches to quantitative analysis of gene-expression patterns observed by use of two-dimensional gel electrophoresis. *Clin. Chem.* **30**, 2031–2036.
2. Horgan, G. W. and Glasbey, C. A. (1995) Uses of digital image analysis in electrophoresis. *Electrophoresis* **16**, 298–305.
3. Wilkins, M. R., Pasquali, C., Appel, R. D., Ou, K., Golaz, O., Sanchez, J-C., et al. (1996) From proteins to proteomes: large-scale protein identification by two-dimensional electrophoresis and amino acid analysis. *Bio/Technology* **14**, 61–65.
4. Wilkins, M. R., Williams, K. L., Appel, R. D., and Hochstrasser, D. F. (eds.) (1997). *Proteome Research: New Frontiers in Functional Genomics.* Springer Verlag, New York.
5. Page, M. J., Amess, B., Rohlff, C., Stubberfield, C., and Parekh, R. (1999) Proteomics: a major new technology for the drug discovery process. *Drug Discovery Today* **4**, 55–62.
6. van Holde, K. E. (1985) Physical Biochemistry, 2nd ed. Prentice Hall, Englewood Cliffs, NJ.
7. Tiselius, A. W. K. (1937) A new apparatus for electrophoretic analysis of colloidal mixtures. *Trans. Faraday Soc.* **33**, 524.
8. Burtis, C. A. and Ashwood, E. R. (1999) Tietz Textbook of Clinical Chemistry, 3rd ed. WB Saunders, Philadelphia.
9. Kenrick, K. G. and Margolis, J. (1970) Isoelectric focusing and gradient gel electrophoresis: a two-dimensional technique. *Analyt. Biochem.* **33**, 204–207.
10. Klose, J. (1975) Protein mapping by combined isoelectric focusing and electrophoresis in mouse tissues: a novel approach to testing for induced point mutations in mammals. *Humangenetik* **26**, 231–234.
11. O'Farrell, P. H. (1975) High resolution two-dimensional electrophoresis of proteins. *J. Biol. Chem.* **250**, 4007–4021.
12. Scheele, G. A. (1975) Two-dimensional gel analysis of soluble proteins: characterisations of guinea pig exocrine pancreatic proteins. *J. Biol. Chem.* **250**, 5375–5385.

13. Schägger, H. and von Jagow, G. (1987) Tricine-sodium dodecyl sulfate-polyacrylamide gel electrophoresis for the separation of proteins in the range from 1 to 100 kDa. *Analyt. Biochem.* **166**, 368–379.

14. Wilkins, M. R., Sanchez, J.-C., Gooley, A. A., Appel, R. D., Humphrey-Smith, I., Hochstrasser, D. F., and Williams, K. L.(1995) Progress with proteome projects: why all proteins expressed by a genome should be identified and how to do it. Biotechnol. *Genet. Engng. Rev.* **13**, 19–50.

15. Bjellqvist, B., Ek, K., Righetti, P. G., Gianazza, E., Görg, A., Westermeier, R., and Postel, W. (1982) Isoelectric focusing in immobilized pH gradients: principle, methodology and some applications. *J. Biochem. Biophys. Methods* **6**, 317–339.

16. Hochstrasser, D. F., Frutiger, S., Paquet, N., Bairoch, A., Ravier, F., Pasquali, C., et al. (1992) Human liver protein map: A reference database established by microsequencing and gel comparison. *Electrophoresis* **13**, 992–1001.

17. Laemmli, U. K. (1970) Cleavage of structural proteins during the assembly of the head of bacteriophage T4. *Nature* **277**, 680–685.

18. Appel, R., Hochstrasser, D. F., Roch, C., Funk, M., Muller, A. F., and Pellegrini, C. (1988) Automatic classification of two-dimensional gel electrophoresis pictures by heuristic clustering analysis: a step toward machine learning. *Electrophoresis* **9**, 136–142.

19. Appel, R., Hochstrasser, D. F., Funk, M., Vargas, J. R., Pellegrini, C., Muller, A. F., and Scherrer, J. R. (1991) The MELANIE project: from a biopsy to automatic protein map interpretation by computer. *Electrophoresis* **12**, 722–735.

20. Appel, R., Palagi, P. M., Walther, D., Vargas, J. R., Sanchez, J. C., Ravier, F., et al. (1997) MELANIE II — A third-generation software package for analysis of two-dimensional electrophoresis images: I. Features and user interface. *Electrophoresis* **18**, 2724–2734.

21. Vincens, P. (1986) HERMeS: a second generation approach to the automatic analysis of two-dimensional electrophoresis gels. Part II: Spot detection and integration. *Electrophoresis* **7**, 357–367.

22. Vincens, P., Paris, N., Pujol, J. L., Gaboriaud, C., Rabilloud, T., Pennetier, J. L., et al. (1986) HERMeS: a second generation approach to the automatic analysis of two-dimensional electrophoresis gels. Part I: Data acquisition. *Electrophoresis* **7**, 347–356.

23. Vincens, P. and Tarroux, P. (1987) HERMeS: a second generation approach to the automatic analysis of two-dimensional electrophoresis gels. Part III: Spot list matching. *Electrophoresis* **8**, 100–107.

24. Vincens, P. and Tarroux, P. (1987) HERMeS: a second generation approach to the automatic analysis of two-dimensional electrophoresis gels. Part IV: Data base organization and management. *Electrophoresis* **8**, 173–186.

25. Tarroux, P., Vincens, P., and Rabilloud, T. (1987) HERMeS: a second generation approach to the automatic analysis of two-dimensional electrophoresis gels. Part V: Data analysis. *Electrophoresis* **8**, 187–199.

26. Miller, M. J., Vo, P. K., Nielsen, C., Geiduschek, E. P., and Xuong, N. H. (1982) Computer analysis of two-dimensional gels: semi-automatic matching. *Clin. Chem.* **28**, 867–875.

27. Skolnick, M. M., Sternberg, S. R., and Neel, J. V. (1982) Computer programs for adapting two-dimensional gels to the study of mutation. *Clin. Chem.* **28**, 969–978.
28. Vo, K. P., Miller, M. J., Geiduschek, E. P., Nielsen, C., Olson, A., and Xuong, N. H. (1981) Computer analysis of two-dimensional gels. *Analyt. Biochem.* **112**, 258–271.
29. Vincens, P. (1993) Morphological grayscale reconstruction in image analysis. IEEE Trans. *Image Proc.* **2**, 176–201.
30. Lutin, K. W. A., Kyle, C. F., and Freeman, J. A. (1978) Quantitation of brain proteins by computer-analyzed two dimensions electrophoresis, in *Electrophoresis '78* (Catsimpoolas, ed.), *Developments in Biochemistry*, vol. 2, Elsevier, NY, pp. 93–106.
31. Garrels, J. (1979) Two-dimensional gel electrophoresis and computer analysis of proteins synthesized by clonal cell lines. *J. Biol. Chem.* **254**, 7961–7977.
32. Taylor, J., Anderson, N. L., and Anderson, N. G. (1981) A computerized system for matching and stretching two-dimensional gel patterns represented by parameter lists, in *Electrophoresis '81* (Allen, R. A. and Arnoud, P., eds.), W de Gruyter, NY, pp. 383–400.
33. Tarroux, P. (1983) Analysis of protein patterns during differentiation using 2-D electrophoresis and computer multidimensional classification. *Electrophoresis* **4**, 63–70.
34. Daubechies, I. (1992) Ten Lectures on Wavelets. Society for Industrial and Applied Mathematics, Philadelphia, PA.
35. S-Plus (2000) Data Analysis Products Division, MathSoft, Seattle, WA.
36. Donoho, D. L. and Johnstone, I. M. (1994) Ideal spatial adaptation via wavelet shrinkage. *Biometrika* **81**, 425–455.
37. Donoho, D. L. and Johnstone, I. M. (1995) Adapting to unknown smoothness via wavelet shrinkage. *J. Am. Statist. Assoc.* **90**, 1200–1224.
38. Donoho, D. L., Johnstone, I. M., Kerkyacharian, G., and Picard, D. (1995) Wavelet shrinkage: asymptopia? *J. Roy. Statist. Soc. Ser. B* **57**, 301–369.

5

Statistical Methods for Assessing Biomarkers

Stephen W. Looney

1. Introduction

According to the *Dictionary of Epidemiology*, a *biomarker* is "a cellular or molecular indicator of exposure, health effects, or susceptibility" (*1*, p. 17). In this chapter, the primary focus is on markers of exposure, although the techniques described here can be applied to any type of biomarker. The process of assessing the quality of a biomarker consists of determining if the biomarker has adequate reliability and adequate validity. *Reliability* refers to "the degree to which the results obtained by a measurement procedure can be replicated" (*1*, p. 145). (Reliability is often used interchangeably with the terms *repeatability* and *reproducibility*.) The reliability of a measurement process is most often described in terms of intrarater and interrater reliability. *Intrarater reliability* (sometimes called *intraobserver agreement*) refers to the agreement between two different determinations made by the same individual and *interrater reliability* (sometimes called *interobserver agreement*) refers to the agreement between the determinations made by two different individuals. A reliable biomarker must exhibit adequate levels of both types of reliability. Also of concern in the assessment of the reliability of a biomarker are intersubject, intrasubject, and analytical measurement variability (*2*). The reliability of a biomarker must be established before validity can be examined; if the biomarker cannot be assumed to provide an equivalent result upon repeated determinations on the same biological material, it will not be useful for practical application.

The *validity* of a biomarker is defined to be the extent to which it measures what it is intended to measure. For example, Qiao et al. (*3*) proposed that the expression of a tumor-associated antigen by exfoliated sputum epithelial cells

From: *Methods in Molecular Biology, vol. 184: Biostatistical Methods*
Edited by: S. W. Looney © Humana Press Inc., Totowa, NJ

could be used as a biomarker in the detection of preclinical, localized lung cancer. For their biomarker to be valid, there must be close agreement between the classification of a patient (cancer/no cancer) using the biomarker and the diagnosis of lung cancer using the gold standard (in this case, consensus diagnosis using "best information"). As another example, body-fluid levels of cotinine have been proposed for use as biomarkers of environmental tobacco smoke exposure *(4)*. For cotinine level to be a valid biomarker of tobacco exposure, it must be the case that high levels of cotinine correspond to high levels of tobacco exposure and low levels of cotinine correspond to low levels of exposure.

Both reliability and validity have to do with interchangeability. Adequate intrarater reliability means that there is minimal within-rater variability so that regardless of when the analyst performs the biomarker determination, we can safely assume that he or she will produce an equivalent result. Adequate interrater reliability means that there is minimal between-rater variability so that regardless of which analyst performs the biomarker determination, we can safely assume that equivalent results will obtain. Adequate validity means that the biomarker determination can be substituted for the gold standard result (assuming that there is a gold standard) or for the standard test result if there is no gold standard.

The appropriate statistical methods for assessing the reliability and validity of a biomarker depend upon the level of measurement of the biomarker. In this chapter, we offer separate recommendations for dichotomous and continuous biomarkers.

2. Dichotomous Biomarkers

2.1. Assessing Reliability of a Dichotomous Biomarker

The same statistical methodology is applied when examining the intra- and interrater reliability of a dichotomous biomarker. Both involve measuring the agreement between two different determinations of the biomarker status of an individual. To assess intrarater reliability, the same analyst would make the determination using the same specimen of material under "identical" conditions. This determination must be blinded, of course, so that the analyst is unaware on the second occasion that he or she is examining the same experimental material that he or she examined on the first occasion. To assess interrater reliability, two different analysts would make the determination using the same specimen of material under "identical" conditions. This determination should also be blinded so that Analyst A is unaware of the result of Analyst B and vice versa. For both intra- and interrater reliability, a 2×2 table is used to show the agreement (and disagreement) between the two determinations.

To assess intrarater reliability, the 2×2 table given in **Table 1** is constructed. To assess interrater reliability, a similar 2×2 table is constructed to show the agreement (and disagreement) between the two determinations made by different individuals on the same biological specimen.

Table 1
2 × 2 Table Showing Agreement Between
Two Determinations by the Same Analyst of the
Same Biological Specimen (Intrarater Reliability)

	Determination 2		
Determination 1	Positive	Negative	Total
Positive	a	b	f_1
Negative	c	d	f_2
Total	g_1	g_2	n

Table 2
2 × 2 Table Showing Agreement
Between a Pathologist and a Cytotechnologist
When Scoring the Same Stained Specimen

	Cytotechnologist		
Pathologist	Positive	Negative	Total
Positive	31	1	32
Negative	0	91	91
Total	31	92	123

Adapted, with permission, from Table 4 of Tockman et al. *(5)*.

For example, Tockman et al. *(5)* examined the use of murine monoclonal antibodies to a glycolipid antigen of human lung cancer as a biomarker in the detection of early lung cancer. As part of their assessment of the interrater reliability of scoring stained specimens, they compared the results obtained on 123 slides read by both a pathologist and a cytotechnologist. They obtained the results given in **Table 2**.

Once the appropriate 2 × 2 table has been constructed, it is desirable to calculate a single numerical quantity as a measure of the reliability of the biomarker. The two most commonly used measures of agreement between two dichotomous variables are the *Index of Crude Agreement*, given by

$$p_0 = (a + d) / n, \tag{1.1}$$

and *Cohen's kappa*, given by

$$\kappa = (p_0 - p_e) / (1 - p_e),$$

where p_e = the percentage agreement between methods A and B that "can be attributed to chance" *(6)*. The estimated percentage agreement between methods A and B that can be attributed to chance is given by

$$p_e = p_1 p_2 + q_1 q_2,$$

where $p_1 = (a+b)/n$, $p_2 = (a+c)/n$, $q_1 = 1-p_1$, and $q_2 = 1-p_2$. The formula for Cohen's kappa now becomes

$$\kappa = \frac{2(ad - bc)}{n^2\,(p_1q_2 + p_2q_1)}\;.\tag{1.2}$$

For the data given in **Table 2**, we obtain $\kappa = 0.979$ using **Eq. (1.2)**. This indicates excellent interrater reliability.

Kappa has the value 1 if there is perfect agreement ($b=c=0$), the value -1 if there is perfect disagreement ($a = d = 0$), and the value 0 if $p_0 = p_e$. Landis and Koch (**7**, p. 165) provide the following guidelines for interpreting the magnitude of kappa:

Value of κ	Interpretation
< 0.00	Poor
0.00 – 0.20	Slight
0.21 – 0.40	Fair
0.41 – 0.60	Moderate
0.61 – 0.80	Substantial
0.81 – 1.00	Almost perfect

Cohen's kappa is the generally accepted method for assessing agreement between two dichotomous variables, neither of which can be assumed to be the gold standard (**8**), but several deficiencies have been noted [(**9**, p. 545), (**10**, p. 425)]. These deficiencies include: (1) If either method classifies no subjects into one of the two categories, $\kappa = 0$. (2) If there are no agreements for one of the two categories, $\kappa < 0$. (3) The value of κ is affected by the difference in the relative frequency of "disease" and "no disease" in the sample. The higher the discrepancy, the larger the value of p_e and the smaller the value of κ. (4) The value of κ is affected by any discrepancy between the relative frequency of "disease" for method A and the relative frequency of "disease" for method B. The greater the discrepancy, the smaller the expected agreement, and the larger the value of κ.

To adjust for these deficiencies, Byrt et al. (**10**) propose that, in addition to κ, one also report the prevalence-adjusted and bias-adjusted kappa (PABAK),

$$PABAK = \frac{(a+d) - (b+c)}{n} = 2p_0 - 1$$

where p_0 is the index of crude agreement given in **Eq. (1.1)**. (Note that PABAK is equivalent to the proportion of "agreements" between the two variables minus the proportion of "disagreements.")

As an illustration of some of the deficiencies of κ, consider the hypothetical data on the agreement between two observers given in **Table 3**.

Table 3
Hypothetical 2 × 2 Table Showing
Agreement Between Two Observers

	Observer B		
Observer A	Positive	Negative	Total
Positive	80	15	95
Negative	5	0	5
Total	85	15	100

Even though the two observers agree on 80% of the specimens, the value of κ is –0.08, indicating poor agreement *(7)*. Two of the previously mentioned deficiencies are at work here. First, because the two "observers" did not agree on any of the subjects who were classified as "negative," $\kappa < 0$. Second, the value of κ is adversely affected by the difference in the average relative frequencies of "disease" (90%) and "no disease" (10%) in the sample. The PABAK coefficient, which adjusts for both of these shortcomings, has the value $2p_0 - 1 = 2(0.80) - 1 = 0.60$. This is considered "moderate" agreement by the Landis and Koch criteria *(7)* and is a much more accurate representation than κ of the agreement between the two observers suggested by **Table 3**.

In additional to using κ and the PABAK coefficient to measure overall agreement, it is also advisable to describe the agreement separately in terms of those specimens that appear to be positive and those that appear to be negative. Using measures of positive agreement and negative agreement in assessing reliability is analogous to using sensitivity and specificity in assessing validity in the presence of a gold standard (*see* **Subheading 2.2.1.**). Such measures can be used to help diagnose the type(s) of disagreement that may be present.

Cicchetti and Feinstein *(11)* proposed indices of *average positive agreement* (p_{pos}) and *average negative agreement* (p_{neg}) for this purpose:

$$p_{pos} = \frac{a}{(f_1 + g_1)/2} \tag{1.3}$$

$$p_{neg} = \frac{d}{(f_2 + g_2)/2}.$$

Note that the denominators of p_{pos} and p_{neg} are the average number of subjects that the two methods classify as positive and negative, respectively. For the data in **Table 3**, $p_{pos} = 2(80)/(95 + 85) = 88.9\%$ and $p_{neg} = 2(0)/(5 + 15) = 0.0\%$. Thus, there is moderate overall agreement between the two observers (as measured by the PABAK coefficient of 0.60), "almost perfect agreement" on specimens

Table 4
"Truth Table" for Tumor-Associated
Antigen as a Biomarker for Lung Cancer

Biomarker	Gold standard		Total
	Positive	Negative	
Positive	42	23	65
Negative	15	53	68
Total	57	76	133

Adapted, with permission, from Table 3 of Qiao et al. *(3)*.

that appear to be positive, and no agreement on specimens that appear to be negative. Thus, efforts to improve the biomarker determination process should be targeted toward those specimens that are negative. (Several computationally intensive methods for estimating sensitivity and specificity in the absence of a gold standard have been proposed [e.g., *12–14*]; however, these are beyond the scope of this chapter.)

2.2. Assessing Validity of a Dichotomous Biomarker
2.2.1. Gold Standard Is Available

Just as in the assessment of reliability described in the preceding, the assessment of the validity of a dichotomous biomarker involves the use of a 2×2 table. If a gold standard is available for the exposure or outcome that the biomarker is intended to represent (the "event"), then the term *conformity* is used to describe the agreement between the biomarker and the occurrence of the event and the term *truth table* is used to describe the 2×2 table.

For example, Qiao et al. *(3)* examined the agreement between the biomarker proposed by Tockman et al. *(5)* and the gold standard method for diagnosing lung cancer. The truth table for their data is given in **Table 4**.

The three measures of conformity obtained from this table are (1) *sensitivity* = a /(a + c) = 42/57 = 73.7%, the percentage of those that experienced the event that the biomarker correctly identified; (2) sp*ecificity* = d/(b + d) = 53/76 = 69.7%, the percentage of those that did not experience the event that the biomarker correctly identified; and *(3) accuracy* = (a + d)/n = (42 + 53)/133 = 71.4%, the percentage of all subjects that the biomarker correctly identified. Qiao et al. *(3)* compared these results with the "standard methods" of chest X-ray and sputum cytology and found that the biomarker proposed by Tochman et al. *(5)* had higher sensitivity, lower specificity, and slightly higher accuracy than both of the standard methods.

Table 5
Hypothetical 2 × 2 Table for
Comparison of Two Biomarkers for Lung Cancer

	Sputum cytology		
Immunocytochemistry	Positive	Negative	Total
Positive	12	53	65
Negative	0	68	68
Total	12	121	133

2.2.2. Gold Standard Is Not Available

If a gold standard is not available for the exposure or outcome that the biomarker is intended to represent (the "event"), then the term *consistency* is used to describe the agreement between the biomarker and some other method used to determine if the event has occurred. This "other method" may be the "standard method" or a competing biomarker. The methods used for assessing intra- and interrater reliability described in **Subheading 2.1.** can be used to assess validity in this situation.

For example, suppose that no gold standard had been available in the study by Qiao et al. *(3)* referred to earlier. Then the investigators could have compared their biomarker based on immunocytochemistry with two "standard" methods of detecting preclinical, localized lung cancer (chest X-ray and sputum cytology). A hypothetical 2 × 2 table for the comparison of their biomarker with sputum cytology based on the assumption that their biomarker agreed with the sputum cytology result on all positive cases of the disease is given in **Table 5**.

Even though the two methods agree on 60% of the specimens, the value of κ is only 0.188, indicating slight agreement *(7)*. Using the PABAK coefficient provides little improvement: PABAK=$2p_0$–1=2(0.602)–1=0.203. The indices of positive and negative agreement are $p_{pos} = 2(12)/(12 + 65) = 31.2\%$ and $p_{neg} = 2(68)/(121 + 68) = 72.0\%$, respectively. Thus, the disagreement between the two methods can be attributed primarily to those specimens that are thought to be positive. A similar analysis for the comparison of the biomarker based on immunocytochemistry with chest X-ray yields κ = 0.483 and PABAK = 0.489, indicating only moderate agreement (data not shown). The indices of positive and negative agreement are $p_{pos} = 64.6\%$ and $p_{neg} = 80.0\%$, indicating once again that the disagreement between the two methods can be attributed primarily to those specimens that are thought to be positive.

Table 6
Hypothetical Data on the Agreement
Between Measurements A and B

Specimen number	Measurement A	Measurement B
1	31	206
2	4	28
3	17	112
4	14	98
5	16	104
6	7	47
7	11	73
8	4	43
9	14	93
10	7	57
11	10	87

3. Continuous Biomarkers

3.1. Use of Pearson's Correlation Coefficient

The most commonly used method for measuring agreement between two continuous variables X and Y is Pearson's correlation coefficient (PCC), denoted by r. However, at least as far back as 1973, it was recognized that the PCC is not appropriate for this purpose *(15)*. The PCC measures *strength of linear association* between two variables, not agreement. We have perfect *agreement* between X and Y if and only if all points in a scatterplot of Y vs X lie along the line $Y = X$; however, we have perfect *correlation* between Y and X if all points in the scatterplot lie along *any* straight line. There are several other shortcomings of the PCC as a measurement of agreement; for example, it is dependent on the heterogeneity of the sample measurements and it is not related to the scale of measurement or to the size of error that might be clinically allowable (*see* [*15,16*] for further discussion).

To illustrate how the PCC can be misleading as a measure of agreement, consider the hypothetical data presented in **Table 6**. A useful first step in assessing *agreement* between X and Y is to construct the scatterplot and then superimpose the line $Y = X$ to get an idea of the deviation of the agreement between X and Y from 1.0. The data in **Table 6** are displayed in this manner in **Fig. 1**. The PCC between measurements A and B is almost perfect, $r = 0.989$, yet there is an obvious deviation from perfect agreement, with the value for measurement A being consistently less than the corresponding value for

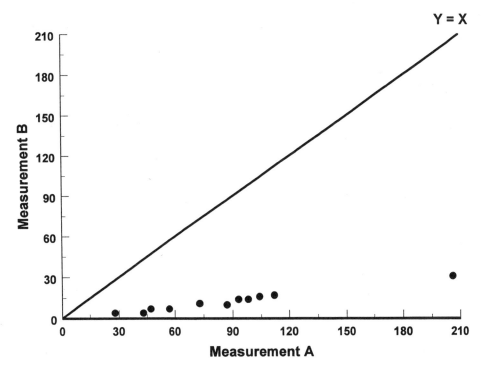

Fig. 1. Scatterplot of hypothetical data on agreement between biomarkers *A* and *B* with the line of perfect agreement (*Y* = *X*) superimposed.

measurement *B*. In the subheadings that follow, we describe alternatives to the PCC for measuring agreement and make recommendations for their appropriate use.

3.1.1. Assessing Intra- and Interrater Reliability

An alternative to the PCC that has been recommended for use in measuring agreement between two continuous measurements, neither of which is the gold standard, is the *intraclass correlation coefficient* (ICC), denoted by r_I *(17)*. To assess both intra- and interrater reliability, a biomarker determination will be made repeatedly for each of n specimens. For intrarater reliability, the same specimen will typically be analyzed on two separate occasions by the same observer. For interrater reliability, the same specimen will typically be analyzed by two different observers. The ICC measures the size of the within-specimen variability relative to the between-specimen variability. It ranges between a value of 0, with $r_I = 0$ indicating no reproducibility at all (large within-specimen variability and zero between-specimen variability), and a value of 1, with $r_I = 1$ indicating perfect reproducibility (large between-specimen variability and zero within-specimen variability). Fleiss *(18)* provided guidelines for interpreting the magnitude of r_I:

Value of r_I	Interpretation
<0.40	Poor
0.40 – 0.75	Fair to good
0.75 – 1.00	Excellent

Suppose that n specimens are each repeatedly analyzed m times (replicates). (Typically, $m = 2$ for both intra- and interrater reliability.) We will assume that the n specimens constitute a random sample. We will also assume that the m replicates (the *occasions* at which the biomarker determinations are made in the case of intrarater reliability and the *observers* in the case of interrater reliability) also constitute a random sample. The simplest method for calculating the appropriate ICC under these assumptions is to use the two-way random effects model without interaction:

$$Y_{ij} = \mu + \alpha_i + \beta_j + \varepsilon_{ij}; \qquad 1 \le i \le n; \qquad 1 \le j \le m$$

where

$$\begin{aligned}
Y_{ij} &= \quad \text{biomarker value for specimen i and replicate } j \\
\mu &= \quad \text{population mean response} \\
\alpha_i &= \quad \text{offset in mean response for specimen } i \\
\beta_j &= \quad \text{offset in mean response for replicate } j \\
\varepsilon_{ij} &= \quad \text{biomarker measurement error}
\end{aligned}$$

The assumptions that underlie this model are as follows: $\alpha_i \sim N(0, \sigma_\alpha^2)$, $\beta_j \sim N(0, \sigma_\beta^2)$, $\varepsilon_{ij} \sim N(0, \sigma_\varepsilon^2)$, where $N(\theta, \delta^2)$ denotes the normal (Gaussian) distribution with mean θ and variance δ^2. All of these assumptions taken together imply that $Y_{ij} \sim N(\mu, \sigma^2)$, where $\sigma^2 = \sigma_\alpha^2 + \sigma_\beta^2 + \sigma_\varepsilon^2$. The population value of the ICC is defined to be $\rho_I = \sigma_\alpha^2/\sigma^2$ and the sample value r_I is obtained by

$$r_I = \max\left[0, \widehat{\sigma}_\alpha^2 / \widehat{\sigma}^2\right]$$

where $\widehat{\sigma}_\alpha^2$ and $\widehat{\sigma}^2$ are the sample estimates of the variance components from the two-way random effects model. After some simplification,

$$r_I = \frac{MSS - MSE}{MSS + (m-1)MSE + m(MSR - MSE)/n} \qquad (3.1)$$

where *MSE* denotes the mean square due to error, *MSS* denotes the mean square due to specimens, *MSR* denotes the mean square due to replicates, m denotes the number of replicates for each specimen, and n denotes the number of specimens. The formulas for calculating these mean squares are as follows:

$$MSS = m \sum_{i=1}^{n} \left(\bar{y}_{i.} - \bar{\bar{y}} \right)^2 / (n-1) ,$$

$$MSR = n \sum_{j=1}^{m} \left(\bar{y}_{.j} - \bar{\bar{y}} \right)^2 / (m-1) , \qquad (3.2)$$

$$MSE = \left(\sum_{i=1}^{n} \sum_{j=1}^{m} \left(y_{ij} - \bar{\bar{y}} \right)^2 - (n-1) MSB - (m-1) MSR \right) / (nm - n - m + 1)$$

where

$$\bar{y}_{i.} = \sum_{j=1}^{m} y_{ij} / m, \; \bar{y}_{.j} = \sum_{i=1}^{n} y_{ij} / n, \; \bar{\bar{y}} = \sum_{i=1}^{n} \bar{y}_{i.} / n .$$

As an example, consider the data in **Table 7,** taken from a study of a bile-acid- induced apoptosis assay for colon cancer risk *(19)*. This is an example of evaluating interrater reliability with $n = 15$ and $m = 2$. Applying the formulas in **Eqs. (3.1)** and **(3.2)**, we obtain $MSS = 698.919$, $MSR = 246.533$, and $MSE = 43.176$ and

$$r_I = \frac{698.919 - 43.176}{698.919 = (2-1)\,(43.176) + 2(246.533 - 43.176)\,/15} = 0.8525 ,$$

which indicates excellent interrater reliability *(18)*.

An alternative to the ICC that is useful in evaluating the intra- and interrater reliability of biomarkers is *Lin's coefficient of concordance (20)*, defined in the population to be

$$\rho_c = 1 - \frac{E\left[(X_1 - X_2)^2 \right]}{\sigma_1^2 + \sigma_2^2 + (\mu_1 - \mu_2)^2} ,$$

where

$\mu_1 = $ mean of X_1
$\mu_2 = $ mean of X_2
$\sigma_1^2 = $ variance of X_1
$\sigma_2^2 = $ variance of X_2

or, in the case $\sigma_1^2 = \sigma_2^2$

$$\rho_c = \frac{\rho}{1 + \left(\dfrac{\mu_2 - \mu_2}{\sigma \sqrt{2}} \right)^2} .$$

The corresponding sample quantity, r_c , is

$$r_c = \frac{2s_{12}}{s_1^2 + s_2^2 + (\bar{x}_1 - \bar{x}_2)^2}$$

Table 7
Interrater Reliability Data from a Study of a
Bile-Induced Apoptosis Assay for Colon Cancer Risk (19)

Specimen	Observer 1	Observer 2
1	11	27
2	9	15
3	54	72
4	55	63
5	50	65
6	44	49
7	58	51
8	5	8
9	21	30
10	58	43
11	41	40
12	59	62
13	39	52
14	34	49
15	23	21

Data courtesy of Carol Bernstein, personal communication, October 17, 2000.

where

s_{12} = sample covariance of X_1 and X_2

$\bar{x}_1$ = sample mean of X_1

$\bar{x}_2$ = sample mean of X_2

s_1^2 = sample variance of X_1

s_2^2 = sample variance of X_2.

It can be shown that $r_c = 1$ if there is perfect agreement between the sample values of X_1 and X_2, $r_c = -1$ if there is perfect negative agreement, and $-1 < r_c < 1$ otherwise. The interpretation of the value of Lin's coefficient is the same as that for the ICC given earlier. The calculation of r_c for the data given in **Table 7** proceeds as follows:

$$\bar{x}_1 = 37.40, \bar{x}_2 = 43.13, s_1^2 = 368.543, s_2^2 = 373.552, s_{12} = 327.9990 .$$

Therefore,

$$r_c = \frac{2s_{12}}{s_1^2 + s_2^2 + (\bar{x}_2 - \bar{x}_2)^2} = \frac{2(327.999)}{368.544 + 373.5523 + (37.4 - 43.13)^2} = 8.847 ,$$

an almost identical result to the ICC of 0.853. The PCC for these data is 0.884.

For an example of data that exhibit strong correlation, but poor agreement, again consider the data in **Table 6**. The PCC between X and Y is almost perfect, $r = 0.989$; however, the intraclass correlation is zero, and Lin's coefficient is only 0.102. The PCC indicates near-perfect linear association, but both of the latter coefficients indicate extremely poor agreement, a much more accurate representation of what is indicated by the plot in **Fig. 1**. In **Subheading 3.2.2.2.**, we present a method for the detailed analysis of the disagreement between two measurements.

3.1.2. Assessing Intersubject, Intrasubject, and Analytical Measurement Variability

The approach described in this subheading very closely follows the scheme proposed by Taioli et al. *(2)* for evaluating the reliability of a biomarker. In addition to intra- and interrater reliability, there are three major components of biomarker variability that must be considered when evaluating reliability. These are: *intersubject* variability, sources of which might include genetics, race, gender, diet; *intrasubject* variability, sources of which include random biologic variation, change in diet, change in exposure; and *analytical* or *laboratory* variability, sources of which include variation between analytical batches, variation within analytical batches, and random variation within the measurement process itself. As Taioli et al. *(2*, p. 308) point out, even if a biomarker has acceptable validity, an excess of intraindividual and/or laboratory variability might render it unusable for research purposes.

To examine each of the sources of variability mentioned previously, biological specimens from each of n subjects are analyzed on m occasions (e.g., *weeks*), and the biomarker determination is repeated for each of r replicate samples (e.g., *aliquots*) from each specimen on each occasion. Replicate samples must be used for all aspects of a proper assessment of biomarker reliability to be examined.

The data from these biomarker determinations are used to estimate the components of biomarker variability mentioned previously. The data from multiple subjects are used to assess intersubject variability, data from multiple occasions are used to assess intrasubject variability, and data from replicate samples are used to assess analytical variability. Once an estimate of analytical variability (or "error variance") is available, it can be used in method comparison studies (*see* **Subheading 3.2.2.2.**). The estimates of intersubject variability, intrasubject variability, and analytical variability can also be combined to form an estimate of the total variance of the biomarker determination, which is useful in calculating the appropriate sample size for future studies in which the biomarker will be used *(2)*.

The statistical model that underlies the approach of Taioli et al. *(2)* is very similar to the one used in assessing inter- and intrarater reliability for continuous biomarkers (*see* **Subheading 3.1.1.**), except that now the replicate observations must be accounted for:

$$Y_{ijk} = \mu + \alpha_i + \beta_j + (\alpha\beta)_{ij} + \varepsilon_{ijk} \; ; \; 1 \le i \le n; \; 1 \le j \le m; \; 1 \le k \le r \qquad (3.3)$$

where

Y_{ijk} = biomarker value for subject i on occasion j and replicate k
μ = population mean response
α_i = offset in mean response for subject i
β_j = offset in mean response for occasion j
$(\alpha\beta)_{ij}$ = offset in mean response for the interaction between subject i and occasion j
ε_{ijk} = biomarker measurement error

(A nonzero interaction term indicates that the differences among subjects vary from occasion to occasion.)

The assumptions that underlie model (3.3) are as follows:

$$\alpha_i \sim N(0,\sigma_\alpha^2), \; \beta_i \sim N(0,\sigma_\beta^2), \; (\alpha\beta)_{ij} \sim N(0,\sigma_{\alpha\beta}^2), \text{ and } \varepsilon_{ijk} \sim N(0,\sigma_\varepsilon^2) \; .$$

All of these assumptions taken together imply that

$$Y_{ijk} \sim N(\mu,\sigma^2) \text{, where } \sigma^2 = \sigma_\alpha^2 + \sigma_\beta^2 + \sigma_{\alpha\beta}^2 + \sigma_\varepsilon^2 \; .$$

As in **Subheading 3.1.1.**, the appropriate statistical method for analyzing biomarker data of this type is *two-way random-effects analysis of variance (ANOVA)*. This analysis will yield tests of significance for each main effect (subjects and occasions) and a test of the interaction between subjects and occasions. It will also provide estimates of each variance component (intersubject variability, intrasubject variability, and analytical variability).

Taioli et al. *(2)* provide an example of the application of their approach to assessing the reliability of four different biomarkers for exposure to carcinogenic metals. The biomarkers they examined are (1) DNA–protein crosslink (DNA–PC), (2) DNA–amino acid crosslink (DNA–AA), (3) metallothionein gene expression (MT), and (4) autoantibodies to oxidized DNA bases (DNAox). We consider the results of only one of their studies here (DNA–PC). In this study, weekly blood samples were drawn three times ($m = 3$) from each of five healthy, unexposed subjects ($n = 5$) and each blood sample was divided into either three or four aliquots ($r = 3$ or 4) for analysis. The blood samples were analyzed during the week in which they were drawn. The results of the random-effects ANOVA are given in **Table 8**. The error variance for the DNA–PC determination is estimated to be 0.0317 and the estimated total variance is $(0.0545 + 0.0176 + 0.0110 + 0.0317) = 0.1148$.

Table 8
Random-Effects ANOVA for DNA–Protein Cross-link Data

Variance component	Variance estimate	F (d.f.)	p-value
Week	0.0545	13.68 (2,7)	< 0.010
Between subject	0.0176	3.45 (4,7)	0.073
Week x subjects	0.0110	2.33 (7,40)	0.045
Error	0.0317	—	—

Adapted, with permission, from Table 2 of Taioli et al. *(2)*.

From **Table 8**, one can see that there is a significant "week" effect (i.e., intrasubject variability), and that the "subject" effect (i.e., intersubject variability) does not quite reach statistical significance. There is also a significant interaction between "week" and "subjects"; this suggests that the "week" effect varies across subjects. The authors point out that, by analyzing the blood samples in the week in which they were drawn, they introduced a possible batch effect that is confounded with the "week" effect. Therefore, the significant intrasubject variability could be a result of the batch effect and not of true week-to-week variation. To prevent this batch effect in the future, the authors modified their assay so that the DNA–PC determination could be performed for all samples at one time.

3.2. Assessment of the Validity of a Continuous Biomarker

3.2.1. Gold Standard Is Available

The assessment of the validity of a continuous biomarker in the presence of a gold standard is equivalent to the calibration of the biomarker *(21)* and is beyond the scope of this chapter. Numerous detailed accounts of methods for calibrating a biomarker are already available (e.g., *[22]*).

3.2.2. Gold Standard Is Not Available

This is equivalent to what is commonly referred to as a "method comparison study" *(15,16,23)*. We have already noted the problems with using the PCC for measuring agreement between continuous variables and Westgard and Hunt *(15,* p. 53) go so far as to state that "the correlation coefficient ... is of no practical use in the statistical analysis of comparison data." The ICC, which has been proposed as an alternative to the PCC for measuring agreement between two continuous variables *(24)*, is also not appropriate as a measure of consistency between two different biomarkers, primarily because to use the version of the ICC recommended in *(24)* requires the assumption that the two biomarkers being considered are a random sample from the population of all biomarkers (*25*, p. 338). There are other disadvantages of using the ICC for measuring agreement between two biomarkers, including some of the same disadvantages involved in using the PCC namely, that it is dependent on the heterogeneity of the sample measurements, and it is not related to the scale of measurement or to the size of error that might be clinically allowable (*25,26*). (Lin's coefficient of concordance is also affected by the heterogeneity of the sample *[27]*, but see also *[28]*.) In the next three subheadings, we present alternative methods that have none of the disadvantages of the ICC.

3.2.2.1. THE BLAND–ALTMAN METHOD

An alternative method for measuring consistency between two biomarkers X_1 and X_2 in which both biomarker determinations are in the same units is to apply the methodology proposed by Altman and Bland (*16,21*). The steps involved in this approach are as follows:

1. Construct a scatterplot and superimpose the line $X_2 = X_1$.
2. Plot the difference between X_1 and X_2 (denoted by d) vs the mean of X_1 and X_2 for each subject.
3. Perform a visual check to make sure that the within-subject repeatability is not associated with the size of the measurement, that is, that the bias (as measured by $[X_1 - X_2]$) does not increase (or decrease) systematically as $(X_1 + X_2)/2$ increases.
4. Perform a formal test to confirm the visual check in **step 3** by testing the hypothesis H_0: $\rho = 0$, where ρ = the true correlation between $(X_1 - X_2)$ and $(X_1 + X_2)/2$.
5. If there is no association between the size of the measurement and the bias, then proceed to **step 6** below. If there does appear to be significant association, then an attempt should be made to find a transformation of X_1, X_2, or both so that the transformed data do not exhibit any association. This can be accomplished by repeating **steps 2–4** for the transformed data. The logarithmic transformation has been found to be most useful for this purpose. (If no transformation can be found, Altman and Bland *[16]* recommend describing the differences between the methods by regressing $[X_1 - X_2]$ on $[X_1 + X_2]/2$.)
6. Calculate the "limits of agreement"; $\bar{d} - 2s_d$ to $\bar{d} + 2s_d$, where $\bar{d}$ is the mean difference between X_1 and X_2 and s_d is the standard deviation of the differences.
7. Approximately 95% of the differences should fall within the limits in **step 6** (assuming a normal distribution). If the differences within these limits are not clinically relevant, then the two methods can be used interchangeably. However, it is important to note that this method is applicable *only* if both measurements are made in the same units.

For example, Bartczak et al. (*29*) compared a high-pressure liquid chromatography (HPLC)-based assay and a gas chromatography (GC)-based assay for urinary muconic acid, both of which have been used as biomarkers of exposure to benzene. Their data, after omitting an outlier due to an unresolved chromatogram peak, are given in **Table 9**.

Table 9
Data on Comparison of Determinations of Muconic Acid
in Human Urine by HPLC–Diode Array and GC–MS Analysis

Specimen number	HPLC (X_1)	GC–MS (X_2)	$X_1 - X_2$	$(X_1 + X_2)/2$
1	139	151	−12.00	145.00
2	120	93	27.00	106.50
3	143	145	−2.00	144.00
4	496	443	53.00	469.50
5	149	153	−4.00	151.00
6	52	58	−6.00	55.00
7	184	239	−55.00	211.50
8	190	256	−66.00	223.00
9	32	69	−37.00	50.50
10	312	321	−9.00	316.50
11	19	8	11.00	13.50
12	321	364	−43.00	342.50

Adapted, with permission, from Table 2 of Bartczak et al. *(29)*.

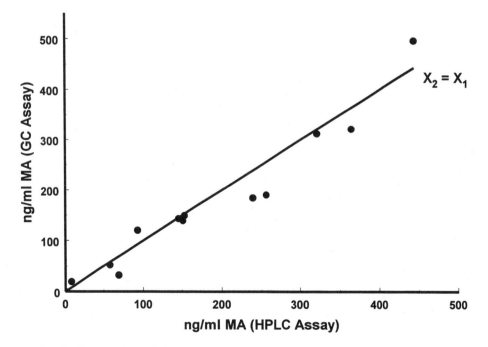

Fig. 2. Scatterplot of data on agreement between (HPLC)-based assay and (GC)-based assay for urinary muconic acid with the line of perfect agreement ($X_2 = X_1$) superimposed.

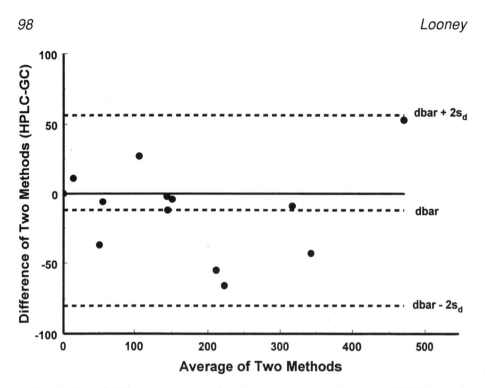

Fig. 3. Plot of difference vs mean for data on agreement between (HPLC)-based assay and (GC)-based assay for urinary muconic acid.

Figure 2 shows the scatterplot of X_2 vs X_1 with the line $X_2 = X_1$ superimposed. This plot indicates fairly good agreement except that 9 of the 12 data points are below the line of agreement. Figure 3 shows the plot of the difference (HPLC – GC) vs the mean of HPLC and GC for each subject. A visual inspection of Fig. 3 suggests that the within-subject repeatability is not associated with the size of the measurement, that is, that (HPLC – GC) does not increase (or decrease) systematically as (HPLC + GC)/2 increases. The sample correlation between (HPLC – GC) and (HPLC + GC)/2 is $r = 0.113$ and the p-value for the test of H_0: $r = 0$ is 0.728. Therefore, the assumption of the independence between the difference and the average is not contradicted by the data. The "limits of agreement" are $\bar{d} - 2s_d = -11.9 - 2(34.2) = -80.3$ to $\bar{d} + 2s_d = -11.9 + 2(34.2) = 56.5$ and these are represented (along with $\bar{d}$) by dotted lines in Fig. 3. (Note that all of the differences fall within the limits $\bar{d} - 2s_d$ to $\bar{d} + 2s_d$.) If differences as large as 80.3 are not clinically relevant, then the two methods can be used interchangeably. Given the order of magnitude of the measurements in Table 9, it would appear that a difference of 80 would be clinically important, so there appears to be inadequate agreement between the two methods. This was not obvious from the plot in Fig. 2.

3.2.2.2. DEMING REGRESSION

Strike *(23)* describes an approach for determining the type of disagreement that may be present when comparing two biomarkers. These methods are most likely to be applicable when one of the methods (method X) is a *reference* method, perhaps a biomarker that is already in routine use, and the other method (method Y) is a *test* method, usually a new biomarker that is being evaluated. Any systematic difference (or *bias*) between the two biomarkers is relative in nature, as neither method can be thought of as representing the true exposure.

As in the Bland–Altman method described in **Subheading 3.2.2.1.**, the first step is to construct a scatterplot of Y vs X and superimpose the line $Y = X$. Any systematic discrepancy between the two biomarkers will be represented on this plot by a general shift in the location of the points away from the line $Y = X$. Strike assumes that systematic differences between the two biomarkers can be attributed to either *constant bias*, *proportional bias*, or both, and assumes the following models for each biomarker result:

$$X_i = \xi_i + \delta_i, \ 1 \leq i \leq n \qquad (3.4)$$
$$Y_i = \eta_i + \varepsilon_i, \ 1 \leq i \leq n$$

where
$$\begin{aligned} X_i &= \text{observed value for biomarker } X, \\ \xi_i &= \text{true value of biomarker } X, \\ \delta_i &= \text{random error for biomarker } X, \\ Y_i &= \text{observed value for biomarker } Y, \\ \eta_i &= \text{true value of biomarker } Y, \\ \varepsilon_i &= \text{random error for biomarker } Y. \end{aligned}$$

Strike further assumes that the errors δ_i and ε_i are stochastically independent of each other and normally distributed with constant variance (σ_δ^2 and σ_ε^2, respectively) throughout the range of biomarker determinations in the study sample. (Strike points out that constant variance assumptions are usually unrealistic in practice and recommends a computationally intensive method for accounting for this lack of homogeneity. This method is incorporated into the MINISNAP software provided with Strike *[23]*).

Strike assumes that any systematic discrepancy between methods X and Y can be represented by

$$\eta_i = \beta_0 + \beta_1 \xi_i \qquad (3.5)$$

In this model, *constant bias* is represented by deviations of β_0 from 0 and *proportional bias* by deviations of β_1 from 1. (This is the same terminology used by Westgard and Hunt *[15]*). If we now incorporate **Eq. (3.5)** into the equation for Y_i in **Eq. (3.4)**, we have

$$Y_i = \beta_0 + \beta_1 X_i + (\varepsilon_i - \beta_1 \delta_i). \qquad (3.6)$$

Model (3.6) is sometimes called a *functional errors-in-variables model* and assessing agreement between biomarkers X and Y requires the estimation of the parameters β_0 and β_1. Strike proposes a method that requires an estimate of the ratio of the error variances given by $\lambda = \sigma_\varepsilon^2/\sigma_\delta^2$. This method is generally referred to in the clinical laboratory literature as "Deming regression"; however, this is somewhat of a misnomer as Deming was concerned with generalizing the errors-in-variables model to nonlinear relationships. Strike points out that the method he advocates for obtaining estimates of β_0 and β_1 is actually due to Kummel *(30)*.

The equations for estimating β_0 and β_1 are as follows:

$$\widehat{\beta_1} = \frac{\left(S_{yy} - \widehat{\lambda}S_{xx}\right) + \sqrt{\left(S_{yy} - \widehat{\lambda}S_{xx}\right)^2 + 4\widehat{\lambda}S_{xy}^2}}{2S_{xy}}, \tag{3.7}$$

$$\widehat{\beta_0} = \overline{Y} - \widehat{\beta_1}\overline{X},$$

$$\widehat{\lambda} = \widehat{\sigma}_\varepsilon^2 / \widehat{\sigma}_\delta^2 ,$$

where

$$S_{yy} = \sum_{i=1}^{n} (y_i - \overline{y})^2, \; S_{xx} = \sum_{i=1}^{n} (x_i - \overline{x})^2, \; S_{xy} = \sum_{i=1}^{n} (x_i - \overline{x})(y_i - \overline{y}) .$$

The estimate $\widehat{\lambda}$ can be obtained either from error variance estimates for each biomarker provided by the laboratory or by estimating each error variance using

$$\widehat{\sigma}^2 = \sum_{i=1}^{n} d_i^2 / (2n)$$

where d_i = difference between the two determinations of the biomarker (replicates) for specimen i. (The error variance can also be estimated from the assessment of reliability recommended by Taioli et al. *[2]* that is described in **Subheading 3.1.2.**) The methodology proposed by Strike cannot be applied without an estimate of the ratio of error variances of the two biomarkers.

To perform significance tests for β_0 and β_1, we need formulas for the standard errors (*SE*s) of $\widehat{\beta}_0$ and $\widehat{\beta}_1$. The approximations that Strike recommends for routine use are given by

$$SE(\widehat{\beta_1}) = \left\{ \frac{\widehat{\beta_1}^2 \left[(1 - r^2)/r^2\right]}{n - 2} \right\}^{1/2}$$

$$SE(\widehat{\beta_0}) = \left\{ \frac{\left[SE(\widehat{\beta_1})\right]^2 \sum X^2}{n} \right\}^{1/2} \tag{3.8}$$

where $$r^2 = [S_{xy}/(S_{xx}S_{yy})^{1/2}]^2$$

is the usual "R^2" value for the regression of Y on X. Tests of $H_0 : \beta_1 = 1$ and $H_0 : \beta_0 = 0$ can be performed by referring $(\hat{\beta}_1 - 1)/SE(\hat{\beta}_1)$ and $(\hat{\beta}_0)/SE(\hat{\beta}_0)$, respectively, to the $t(n-2)$ distribution.

As mentioned earlier, the approach described previously is based on the assumption that the error variances σ_δ^2 and σ_ε^2 are constant throughout the range of biomarker determinations in the study sample. However, as Strike points out, this assumption is usually unrealistic in practice and recommends the "weighted Deming regression" methods of Linnet *(31,32)* for accounting for this lack of homogeneity. These methods are incorporated into the MINISNAP software provided with Strike *(23)*; however, replicate measurements are required for each test specimen using both biomarkers in order to apply these methods.

For example, consider the data in **Table 6** that were discussed in **Subheading 3.1.** The scatterplot of Y vs X in **Fig. 1** indicated substantial lack of agreement between X and Y and this was borne out by the intraclass correlation coefficient and Lin's coefficient, both of which indicated substantial disagreement. We can apply Strike's method to gain a better understanding of this disagreement.

Using the formulas in **Eqs. (3.7)** and **(3.8)**, we obtain $\hat{\beta}_1 = 0.158$, $SE(\hat{\beta}_1) = 0.007$, $\hat{\beta}_0 = -1.342$, $SE(\hat{\beta}_0) = 0.614$. For the test of $H_0 : \beta_1 = 1$, this yields

$$t_{cal} = (\hat{\beta}_1 - 1)/SE(\hat{\beta}_1) = (0.158-1)/0.007 = -129.54,$$

and using a t-distribution with $n - 2 = 9$ degrees of freedom we find $p < 0.0001$. Therefore, there is significant proportional bias (which in this case is negative since $\hat{\beta}_1 < 1.0$). For the test of $H_0 : \beta_0 = 0$, we have

$$t_{cal} = \hat{\beta}_0/SE(\hat{\beta}_0) = -1.342/0.614 = -2.19,$$

and, again using a t-distribution with 9 degrees of freedom, we have $p = 0.056$. Thus, the constant bias is not statistically significant, but just misses the usual cutoff of 0.05.

3.2.2.3. Spearman Correlation

A method that can be used to measure consistency between measurements that are in different units is *Spearman's rank correlation coefficient (SCC)*, denoted by r_s. This method is useful, and to be preferred over Pearson's correlation, when examining the agreement between two biomarkers whose determinations are in different units, or between a biomarker and some other measure of exposure such as environmental monitoring. For example, Coultas et al. *(33)* used the SCC to measure the consistency between various measures of exposure to environmental tobacco smoke at work (e.g., nicotine exposure measured with a personal monitoring pump vs. post-shift urinary cotinine).

The SCC measures the agreement between two sets of measurements, after the measurements have been ordered ("ranked") from smallest to largest. Like other correlation coefficients, the SCC ranges between 1 (perfect agreement) and −1 (perfect negative agreement). As an example, suppose $n = 9$ and that the values obtained from Biomarker A for the nine specimens have been arranged in order from smallest to largest, with the smallest receiving rank 1 and the largest receiving rank 9. The same ordering is repeated for the nine specimens for each of biomarkers B and C. The specimens are arranged in order of their rankings according to biomarker A and then the ranks for biomarkers B and C are also noted:

Rank by biomarker A	1 2 3 4 5 6 7 8 9
Rank by biomarker B	1 2 3 4 5 6 7 8 9
Rank by biomarker C	9 8 7 6 5 4 3 2 1

According to the SCC, biomarkers A and B have perfect agreement ($r_s = 1$), whereas biomarkers A and C have perfect negative agreement ($r_s = -1$).

If r_s is close to 1, then we can assume that a subject with high levels of exposure, according to biomarker A, will also tend to have high levels of exposure, according to biomarker B (and similarly for low levels). Therefore, regardless of which biomarker we use, we can feel confident that subjects will be assigned to a high (or low) exposure group in a consistent manner. However, the PCC is not recommended for this purpose because r may be low even though high levels of exposure, according to biomarker A, are associated with high levels of exposure, according to biomarker B. This can occur, for example, if the relationship between the two biomarkers is nonlinear. Spearman's correlation will have a large value if the two biomarker determinations are strongly related according to *any* monotonic relationship.

The SCC is calculated using the following formula:

$$r_s = 1 - \frac{6 \sum_{i=1}^{n} (r_i - s_i)^2}{n(n^2 - 1)}, \qquad (3.9)$$

where r_i = the rank of subject i according to biomarker A, s_i = the rank of subject i according to biomarker B, and n = the number of subjects. Morton et al. *(34)* provide guidelines that can be used to interpret the value of r_s:

Table 10
Data on Concentrations of *o*-Cresol and Hippuric Acid Concentrations in Urine Samples

Specimen number	*o*-Cresol (μg/mL)	Rank of *o*-cresol	Hippuric acid (mg/mL)	Rank of Hippuric acid
1	0.21	1.5	0.30	2.0
2	0.21	1.5	0.80	5.0
3	0.25	3.0	0.40	3.0
4	0.28	4.0	0.50	4.0
5	0.32	5.0	1.10	7.5
6	0.34	6.0	1.19	9.0
7	0.41	7.0	1.30	12.0
8	0.44	8.5	1.08	6.0
9	0.44	8.5	1.10	7.5
10	0.51	10.0	1.20	10.5
11	0.59	11.0	1.20	10.5
12	0.76	12.0	1.33	13.0
13	1.25	13.0	0.20	1.0
14	1.36	14.0	2.10	14.0
15	2.80	15.0	3.02	15.0

Adapted, with permission, from Tables 1–3 of Amorim and Alvarez-Leite *(35)*.

Value of $|r_s|$	Interpretation
0.00 – 0.20	Negligible
0.21 – 0.50	Weak
0.51 – 0.80	Moderate
0.81 – 1.00	Strong

To illustrate the calculation of the SCC, consider the data in **Table 10** on concentrations of *ortho*-cresol and hippuric acid in urine samples of workers exposed to toluene *(35)*, which have been ranked according to the magnitude of the *o*-cresol value. Applying the formula in Eq. (3.9), we obtain $r_s = 0.632$, which indicates a moderate degree of consistency between the two measurements.

3.2.2.4. CRITERION AND CONSTRUCT VALIDITY

There are two types of validity that should be examined when evaluating a biomarker in the absence of a gold standard. *Criterion validity* is examined by correlating the biomarker with measures of some other phenomenon that is expected to be correlated with the exposure or outcome that the biomarker represents. There are two types of criterion validity, *concurrent* and *predictive*. *Concurrent* refers to other phenomena that are contemporaneous with the biomarker, whereas *predictive* refers to phenomena that occur at some future time point.

For example, to assess concurrent validity in a study of the usefulness of post-shift urinary and salivary cotinine as a biomarker for workplace exposure to environmental tobacco smoke, urinary and salivary cotinine levels of nonsmoking workers were correlated with the total number of smokers and the total number of hours exposed to cigarette smoke in the workplace *(33)*. As an example of predictive validity, in a study of the usefulness of plasma cotinine as a biomarker for environmental tobacco smoke, the authors examined the correlation between plasma cotinine and the metabolic clearance of theophylline, a drug whose metabolism is known to be increased in nonsmokers by the presence of cigarette smoke *(36)*. The PCC is typically used to measure criterion validity; however, we recommend that the SCC be used instead, as the PCC measures only the degree of linear relationship, whereas the SCC is sensitive to any monotonic relationship between the biomarker and the criterion.

The other type of validity that should be evaluated in the absence of a gold standard is *construct validity*, which is examined in light of hypotheses formulated by the investigator about the characteristics of those who should have high levels of the exposure represented by the biomarker vs those who should have low levels. For example, Hüttner et al. *(37)* evaluated chromosomal aberrations in human peripheral blood lymphocytes as a biomarker of chronic exposure to heavy metals and dioxins/furans over a long period of time. As part of their examination of construct validity, they compared 52 exposed individuals from a polluted area with 51 matched controls from a distant nonindustrialized area and found a statistically significant increase in the frequency of chromosomal aberrations in human peripheral blood lymphocytes in the exposed group ($p < 0.001$). Construct validity is generally assessed by performing the appropriate statistical test to carry out the comparison of interest. For example, Hüttner et al. *(37)* used Fisher's exact test to compare the exposed and unexposed individuals in terms of the dichotomous outcome (chromosomal aberration/no chromosomal aberration). For continuous outcomes, the appropriate normal-theory test should be used if the outcome appears to follow a normal distribution (*t*-test for the comparison of two groups, one-way ANOVA for more than two groups). If the outcome data are highly skewed or otherwise non-normal, the Mann–Whitney–Wilcoxon test should be used to compare two groups, and the Kruskal–Wallis test should be used for more than two groups.

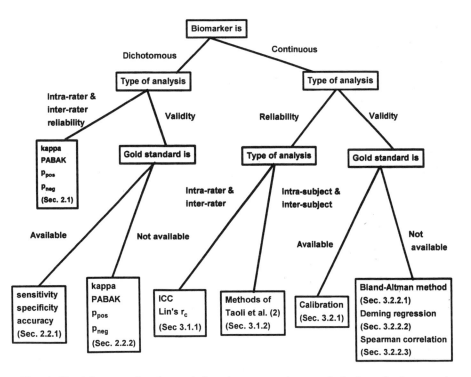

Fig. 4. Decision tree for determining the appropriate statistical method to use in assessing reliability or validity of a biomarker.

4. Discussion

In this chapter, we have described methods for assessing the reliability and validity of biomarkers that we feel are easy to apply and interpret, and whose results can be easily communicated to nonstatisticians. (**Figure 4** summarizes our recommendations in the form of a decision tree for easy reference.) Some of the methods we recommend are controversial; for example, there are those who claim that the adjustment for chance agreement in the calculation of Cohen's kappa is inappropriate when measuring the agreement between clinical observers *(9)* and that the Index of Crude Agreement is the correct measure to use. However, as the use of Cohen's kappa is so widespread and no one has come forward as yet with a convincing argument that kappa should be abandoned entirely, we have chosen to recommend its use, along with the PABAK coefficient and the indices of positive and negative agreement.

Our recommendation against the use of the intraclass correlation coefficient for assessing consistency between competing biomarkers may surprise some readers who are experienced with biomarker evaluation; however, we feel that the criticisms by Bland and Altman *(25)* and Atkinson *(26)* are valid and that the methods recommended by Bland and Altman *(16,21)*, and Strike *(23)* are preferable. Our recommendations are also consistent with the decision tree for the proper use of ICCs presented in *(38)*.

We may also surprise some experienced readers by not recommending that significance tests be performed for κ, PABAK, the intraclass correlation ρ_I, and Spearman's rho. As Kraemer *(39)* and Altman and Bland *(16)* have pointed out, testing H_o: $\kappa = 0$ or H_o: $\rho_I = 0$ is beside the point because it is unlikely that we would be interested in the agreement between totally unrelated quantities in an assessment of reliability or validity. We prefer to use the guidelines provided by various authors as descriptors of the degree of agreement. Of course, these guidelines were not intended to be applicable in every situation and could be modified as necessary for the particular area of study. If a test of significance or confidence interval is required, the software used to calculate the coefficient can be used to produce these results as well (see below for software recommendations).

The coefficient of variation (CV) is commonly used as a measure of variability within assays, between assays, within samples, within individuals, etc. (*see*, e.g., *[35,40]*). However, there are difficulties with the interpretation of the CV as it is usually presented (e.g., "the assay has a CV of 8%"). Strike *(23*, p. 25) provides a very lucid discussion of these difficulties and we agree with his recommendation: "Use the CV if you must, but use it carefully, and with proper qualifications."

In terms of software requirements for carrying out the procedures we have recommended, either the KAPPA or PAIRS programs of the software package PEPI *(41)* can be used to perform any of the calculations described here, with the exception of the intraclass correlation coefficient (ICC) in Eq. (3.1) and Deming Regression. PEPI is very reasonably priced (currently $50) and can be obtained from USD Inc., 2171-F West Park Ct., Stone Mountain, GA 30087, telephone (770) 469-4098, website www.usd-inc.com. The ICC in Eq. (3.1) can be calculated using the SAS code provided in *(24)*, and the MINISNAP software provided with **ref.** *(23)* can be used to perform Deming regression.

Finally, an important issue that we have not addressed in this chapter is the assessment of surrogate markers that are used in place of a dichotomous event (death, recurrence of disease, etc.) as outcomes in clinical trials *(42–45)*. Although many of the methods outlined in this chapter could be used to assess the reliability and validity of a surrogate marker, there are many other issues dealing with the use of these markers that are as yet unresolved, as illustrated by the recent "debate" at a workshop sponsored by the National Institute of Allergy and Infectious Diseases *(45)*. A discussion of these issues is beyond the scope of this chapter. However, Chapter 9 of this text *Statistical Considerations in Assessing Molecular Markers for Cancer Prognosis and Treatment Efficacy* by Dignam et al. does consider some of the issues in detail.

Acknowledgment

I wish to thank Stephen George of the Duke University Medical Center for his many helpful comments that greatly improved this chapter. I also wish to thank Fred Benz of the University of Louisville School of Medicine for his encouragement and his informative presentations.

References

1. Last, J. M. (1995) *A Dictionary of Epidemiology*, (3rd edit.). Oxford University Press, New York.
2. Taioli, E., Kinney, P., Zhitkovich, A., Fulton, H., et al. (1994) Application of reliability models to studies of biomaker validation. *Environ. Health Perspect.* **102**, 306–309.
3. Qiao, Y-L., Tockman, M. S., Li, L., Erozan, Y. S., et al. (1997) A case-cohort study of an early biomarker of lung cancer in a screening cohort of Yunnan tin miners in China. *Cancer Epidemiol. Biomarker Prev.* **6**, 893–900.
4. Benowitz, L. (1999) Biomarkers of environmental tobacco smoke exposure. *Environ. Health Perspect.* **107**(Suppl 2), 349–355.
5. Tockman, M. S., Gupta, P. K., Myers, J. D., Frost, J. K., et al. (1988) Sensitive and specific monoclonal antibody recognition of human lung cancer antigen on preserved sputum cells: a new approach to early lung cancer detection. *J. Clin. Oncol.* **6**, 1685–1693.
6. Cohen, J. (1960) A coefficient of agreement for nominal scales. *Educ. Psychol. Meas.* **20**, 37–46.
7. Landis, J. R. and Koch, G. G. (1977) The measurement of observer agreement for categorical data. *Biometrics* **33**, 159–174.
8. Bartko, J. J. (1991) Measurement and reliability: statistical thinking considerations. *Schizophr. Bull.* **17**, 483–489.
9. Feinstein, A. R. and Cicchetti, D. V. (1990) High agreement but low kappa: I. The problems of two paradoxes. *J. Clin. Epidemiol.* **43**, 543–549.
10. Byrt, T., Bishop, J., and Carlin, J. B. (1993) Bias, prevalence, and kappa. *J. Clin. Epidemiol.* **46**, 423–429.
11. Cicchetti, D. V. and Feinstein, A. R. (1990) High agreement but low kappa: II. Resolving the paradoxes. *J. Clin. Epidemiol.* **43**, 551–558.
12. de Bock, G. H., Houwing-Duistermaat, J. J., Springer, M. P., Kievit, J., and van Houwelingen, J. C. (1994) Sensitivity and specificity of diagnostic tests in acute maxillary sinusitis determined by maximum likelihood in the absence of an external standard. *J. Clin. Epidemiol.* **47**, 1343–1352.
13. Joseph, L., Gyorkos, T. W., and Coupal, L. (1995) Bayesian estimation of disease prevalence and the parameters of diagnostic tests in the absence of a gold standard. *Am. J. Epidemiol.* **141**, 263–272.
14. Hui, S. L. and Zhou, X. H. (1998) Evaluation of diagnostic tests without gold standards. *Statist. Methods Med. Res.* **7**, 354–370.
15. Westgard, J. O. and Hunt, M. R. (1973) Use and interpretation of common statistical tests in method-comparison studies. *Clin. Chem.* **19**, 49–57.

16. Altman, D. G. and Bland, J. M. (1983) Measurement in medicine: the analysis of method comparison studies. *Statistician* **32**, 307–317.

17. Shrout, P. E. and Fleiss, J. L. (1979) Intraclass correlations: uses in assessing rater reliability. *Psychol. Bull.* **86**, 420–428.

18. Fleiss, J. L. (1986) *The Design and Analysis of Clinical Experiments.* John Wiley & Sons, New York.

19. Bernstein, C., Bernstein, H., Garewal, H., Dinning, P., et al. (1999) A bile acid-induced apoptosis assay for colon cancer risk and associated quality control studies. *Cancer Res,* **59**, 2353–2357.

20. Lin, L. I. (1989) A concordance correlation coefficient to evaluate reproducibility. *Biometrics* **45**, 255–268.

21. Bland, J. M. and Altman, D. G. (1986) Statistical methods for assessing agreement between two methods of clinical measurement. *Lancet* Feb. 8, 307–310.

22. Strike, P. W. (1991) *Statistical Methods in Laboratory Medicine.* Butterworth-Heinemann, Oxford.

23. Strike, P. W. (1996) Assay method comparison studies, in *Measurement in Laboratory Medicine: A Primer on Control and Interpretation.* Butterworth-Heinemann, Oxford, pp. 147–172.

24. Lee, J., Koh, D., and Ong, C. N. (1989) Statistical evaluation of agreement between two methods for measuring a quantitative variable. *Comput. Biol. Med.* **19**, 61–70.

25. Bland, J. M. and Altman, D. G. (1990) A note on the use of the intraclass correlation coefficient in the evaluation of agreement between two methods of measurement. *Comput. Biol. Med.* **20**, 337–340.

26. Atkinson, G. (1995) A comparison of statistical methods for assessing measurement repeatability in ergonomics research, in *Sport, Leisure and Ergonomics* (Atkinson, G., and Reilly, T., eds.), E. and F. N. Spon., London, pp. 218–222.

27. Atkinson, G. and Nevill, A. (1997) Comment on the use of concordance correlation to assess the agreement between two variables (Letter to the Editor). *Biometrics* **53**, 775–777.

28. Lin, L. I. and Chinchilli, V. (1997) Rejoinder to the Letter to the Editor from Atkinson and Nevill. *Biometrics* **53**, 777–778.

29. Bartczak, A., Kline, S. A., Yu, R., Weisel, C. P., et al. (1994) Evaluation of assays for the identification and quantitation of muconic acid, a benzene metabolite in human urine. *J. Toxicol. Environ. Health* **42**, 245–258.

30. Kummel, C. H. (1879) Reduction of observation equations which contain more than one observed quantity. *Analyst* **6**, 97–105.

31. Linnet, K. (1990) Estimation of the linear relationship between the measurements of two methods with proportional errors. *Statist. Med.* **9**, 1463–1473.

32. Linnet, K. (1993) Evaluation of regression procedures for methods comparison studies. *Clin. Chem.* **39**, 424–432.

33. Coultas, D. B., Samet, J. M., McCarthy, J. F., and Spengler, J. D. (1990) A personal monitoring study to assess workplace exposure to environmental tobacco smoke. *Am. J. Public Health* **80**, 988–990.

34. Morton, R. F., Hebel, J. R., and McCarter, R. J. (1996) *A Study Guide to Epidemiology and Biostatistics*. Aspen, Gaithersburg, MD.
35. Amorim, L. C. A. and Alvarez-Leite, E. M. (1997) Determination of *o*-cresol by gas chromatography and comparison with hippuric acid levels in urine samples of individuals exposed to toluene. *J. Toxicol. Environ. Health* **50,** 401–407
36. Matsunga, S. K., Plezia, P. M., Karol, M. D., Katz, M. D., et al. (1989) Effects of passive smoking on theophylline clearance. *Clin. Pharmacol. Ther.* **46,** 399–407.
37. Hüttner, E., Götze, A., and Nikolova, T. (1999) Chromosomal aberrations in humans as genetic endpoints to assess the impact of pollution. *Mutat. Res.* **445,** 251–257.
38. Müller, R. and Büttner, P. (1994) A critical discussion of intraclass correlation coefficients. *Statist. Med.* **13,** 2465–2476.
39. Kraemer, H. C. (1980) Extension of the kappa coefficient. *Biometrics* **36,** 207–216.
40. Atawodi, S. E., Lea, S., Nyberg, F., Mukeria, A., et al. (1998) 4-Hydroxyl-1-(3-pyridyl)-1-butanone-hemoglobin adducts as biomarkers of exposure to tobacco smoke: validation of a method to be used in multicenter studies. *Cancer Epidemiol. Biomark. Prev.* **7,** 817–821.
41. Abramson, J. H. and Gahlinger, P. M. (1999) *Computer Programs for Epidemiologists, PEPI Version 3.00*. Brixton Books, Llanidloes, Powys.
42. Fleming, T. R., Prentice, R. L., Pepe, M. S., and Glidden, D. (1994) Surrogate and auxillary endpoints in clinical trials, with potential applications in cancer and AIDS research. *Statist. Med.* **13,** 955–968.
43. Fleming, T. R. (1994) Surrogate markers in AIDS and cancer trials. *Statist. Med.* **13,** 1423–1435.
44. Buyse, M. and Molenberghs, G. (1998) Criteria for the validation of surrogate endpoints in randomized experiments. *Biometrics* **54,** 1014–1029.
45. Albert, J. M., Ioannidis, J. P. A., Reichelderfer, P., Conway, B., et al. (1998) Statistical issues for HIV surrogate endpoints: Point/counterpoint. *Statist. Med.* **17,** 2435–2462.

6

Power and Sample Size Considerations in Molecular Biology

L. Jane Goldsmith

1. Introduction

Sample size is an important interest of researchers in laboratory and clinical settings. The number of cases to be investigated profoundly affects the cost and duration of a study. Sample size is estimated to achieve a certain statistical power and a careful power and sample size analysis can predetermine the success of a study or experiment.

1.1. What Is Statistical Power?

In the hypothesis test setting, statistical power is the probability of rejecting the null hypothesis when it is appropriate to do so, that is, when the null hypothesis is false. It is clear that it is desirable for statistical power to be high, representing a probability close to 1, because power is the probability of drawing the correct conclusion when the null hypothesis is false.

Statistical power is related to β, the probability of an error of type II. β is the probability that the hypothesis test in question will erroneously fail to reject H_0 when H_1, the alternative hypothesis, is true. It is easy to see that:

P[do not reject H_0 | H_1 is true] = β Probability of an error
P[reject H_0 | H_1 is true] = statistical power   Probability of the correct conclusion
         = $1 - \beta$

Statistical power is one of the central concepts in statistics. Many statistical methods are designed to increase power, and new statistical methods are often justified by claims that they are the most powerful or at least enhance power (*1*, pp. 60–63).

From: *Methods in Molecular Biology, vol. 184: Biostatistical Methods*
Edited by: S. W. Looney © Humana Press Inc., Totowa, NJ

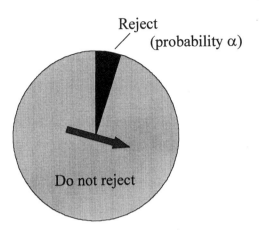

Fig. 1. The hypothesis test spinner.

Indeed, the reduction of the probability of error of type II, or, as outlined previously, the increase of statistical power, is the primary reason for using statistical methods and theory. This can be seen directly by observing that if one wants merely to control α, the probability of rejecting H_0 when it is true, one can merely use a uniform random number generator to generate a random number between 0 and 1. On occasions when the random number is $\leq \alpha$, reject the null hypothesis. Otherwise, do not reject. The probability of a spurious result (the type I error probability) is fixed at α. The statistical power is, unfortunately, also α. This method can also be criticized for the property that it is not based on measurements or data. *See* **Fig. 1** for a prototype random number generator, a hypothesis test spinner with $\alpha = 0.05$.

Using statistical theory, one can design a hypothesis test with the required α and high statistical power to detect whether or not the research hypothesis is true.

1.2. Power and Sample Size

Statistical power and sample size are often discussed in the same breath. Indeed, power increases with sample size in any "consistent" statistical test (*1*, p. 305). The desirable property of consistency in a statistical test is intuitively appealing: the more subjects or cases in the experiment or research study the more likely a correct conclusion. In such a consistent statistical test, "the more the merrier" slogan applies. That is, the more subjects, the more likely the statistical conclusion will be correct.

The hypothesis spinner in **Fig. 1** is an obvious example of a nonconsistent statistical test. No matter how many subjects are included in the experiment, the probability of rejection of the null hypothesis remains constant at α.

1.3. Other Uses for Sample Size Calculations

Sample size calculations are important in other areas of statistical inference, including determination of confidence interval widths and coverage probabilities. These are beyond the scope of the presentation in this chapter.

1.4. Importance in Research

Upon reflection, any scientist or experimenter can see the importance of correct or sufficient statistical power or sample size in an experiment. If the sample size is too small, and, consequently, the statistical power is too low, there is a high probability that the experiment will not detect the effect of interest. That is, even though the research hypothesis is true, the data collected under this experiment design will not yield a statistically significant test result. The researchers involved in such an experiment will be hard pressed to prove that the effect is *not* present, as their research design, because of its low statistical power, was not sensitive enough to detect an effect. In many cases journal editors will not accept a negative (not statistically significant) result for publication if the authors cannot demonstrate adequate sample size and statistical power.

It is easy to see that failure to plan for an experiment with adequate statistical power is risking that a negative conclusion will mean a waste of time and resources, resulting in research that is not meaningful and not publishable.

The convention for planning for statistical power is that an experiment should be designed to have power of 80% or 90%. A recent article uses Bayesian techniques to support different standards for α and β *(2)*.

1.5. Ethics of Power and Sample Size

Many authors have written of the importance of sample size for efficient research *(3–5)*. In biomedical experiments involving increased pain, discomfort, fear, or risk on the part of the subjects, it is apparent that it is entirely unethical to conduct an experiment with too small a sample size. If this is done, the subjects, human volunteers' or helpless animals, have suffered to no avail. Implicit in informed consent is the notion that the research is conducted efficiently. If the sample size and statistical power are wastefully high, then extra subjects have suffered needlessly. In studies involving suffering or sacrifice, the "Goldilocks Principle" applies for sample size: not too big and not too small, but just right.

Even in research where no suffering is involved, resources, money, and the valuable time of researchers and human subjects can be wasted with the wrong sample size.

2. Data Considerations
2.1. Types of Data

In molecular biology and other types of biological research, different types of data may be collected for research purposes. These different data types are summarized below:

1. *Categorical* data are data indicating group membership. Natural examples are allele type, race, country of birth, and religion.

   An important special class of categorical data arises when there are exactly two categories. These data, often characterized as "yes/no" data, are called *dichotomous* data. Natural examples are presence of a trait (yes/no), presence of an allele (yes/ no), death (yes/no), evidence of infection (yes/no), and cure (yes/no).

   Categorical data are often coded as integer values, corresponding to group numbers that are associated with specific groups. Some statistical software packages, such as SAS (*6*), allow character data, or names, to represent categorical data. It is often advisable to code "yes/no," or dichotomous data, as the integers 1 or 0, with the value 1 denoting the occurrence of the event of interest ("yes") and 0 denoting the absence of the event of interest ("no"). This coding scheme sets things up for logistic regression, a statistical method sometimes used to search for predictors of dichotomous events.

2. *Ordinal* data are data reflecting a natural ordering. Ordinal data, however, do not have a "distance" measure associated with them. Examples of ordinal data in medical research include health index measures such as APGAR scores (used to indicate health of neonates), cancer stage (used to indicate the severity or extent of disease in cancer), and anesthesia class (used to indicate general overall health of a surgical patient). In each of these examples, subtraction (computing a "distance" between values) does not make sense. That is, a newborn with an APGAR of 9 is not "2 better" than an infant with an APGAR of 7. It is just known that the baby with APGAR 9 appears healthier at birth than the one with a score of 7, and that a baby with APGAR 8 would rank between them in terms of apparent health.

3. *Numerical* data reflect ordering and distance. Numerical data can be integer data, such as parity (number of live births), number of teeth, number of lesions, and so forth. Medical measurements, such as hematocrit, oxygen saturation, and systolic blood pressure, are examples of numerical data described as *interval-level* data.

2.2. Switching Between Data Types

In biomedical research it is common for data collected naturally or initially as one of the data types described previously to be transformed into data of a different type. Grouping subjects into age-group categories is an example of transformation of numerical data (age) into ordinal data (age group). Statistics programs such as SPSS make this transformation easy (*7*).

Often data are dichotomized, that is, changed from numerical data into high–low categories. "Cut-points" provide the boundaries used to change numerical data into dichotomous data. Biomedical language recognizes this ubiquitous transformation with special nomenclature: for example, the terms *premature*, *hypoglycemic*, and *anemic* all represent dichotomizations of numerical measurements. The description "natural" dichotomous data was used above to indicate that those examples do not represent dichotomizations, but naturally occurring dichotomies of interest.

Another type of transformation involves a progression from dichotomous to numerical data. For example, a sequence of 1 or 0 (yes or no) answers to questions from a survey or checklist can be summed into a numerical score.

Categorical data can sometimes be found through experience and statistical testing to have a natural ordering, thus converting it to ordinal or even numerical type. An example here is the staging system of cancer. Developed as a systematic method of describing the extent of disease at diagnosis, the tumor-node-metastatis (TNM) system has a natural ordering related to the natural progression of the disease, and the stages have been shown to be negatively correlated with survival probability (*8*, pp. 3–5). Ongoing efforts to refine the staging system ascertain that the addition of new stage definitions explain survival meaningfully (*8*, p. 12). Cancer stage at diagnosis is often considered an ordinal or numerical variable.

2.3. Statistical Power, Data Type, and Strategies for Efficiency

Different kinds of statistical methods are appropriate for each data type. Contingency tables, chi-square (χ^2) tests, log-linear models, and logistic regression are among the methods used for categorical and dichotomous data. In many cases, categorical data and associated statistical methods require large sample sizes.

Ordinal data are often analyzed by nonparametric statistical methods. These methods can be very efficient in terms of statistical power and sample size. However, often the most powerful statistical methods are parametric tests on numerical data.

Numerical data represent the most informative data in biomedical research. Statistical theory and experience have indicated that, in general, research utilizing numerical data with parametric statistical methods affords the most efficient statistical power and sample size. Authors recognizing this relationship between the information in data and efficient research deplore the practice of dichotomization or categorization described above as a waste of resources (*9–11*).

In summary, the current best advice is to eschew dichotomization or categorization in data collection. Record and analyze numerical data when possible.

Further advice is to use strategies converting naturally occurring dichotomous or ordinal data to numerical data whenever large samples are not practicable. One such strategy is to sum dichotomous responses, as mentioned previously. Another is to record survival time or time elapsed until the occurrence of an event of interest rather than the simple, relatively uninformative outcome of event/no event. An example here is the notation of age of onset of a disease. Earlier onset may be predicted by a genetic marker, indicating a genetic link to the disorder.

A recent method involves substituting a numerical outcome or "surrogate marker" for dichotomous data. An example is monitoring CEA levels to detect increase in tumor burden rather than monitoring cancer patients for a recurrence of their tumor, a yes/no variate. Recent statistical studies grapple with the problem of determining when surrogate markers are justified *(12–14)*.

3. Experiment Design Considerations

An important concept in understanding statistical power and sample size calculations is that each statistical method has its own associated sample size formula or calculation method. In the early part of the 20th century, when inferential statistics was a new discipline, sample size calculation was rare. In the days of hand-cranked calculators, large samples were a computational burden. "Rule of thumb" methods were often used, with 30 as a popular number *(15)*. As the importance of careful power and sample size determination as part of the research planning process became more recognized, researchers would try to use or adapt simple formulas for more complicated analyses. For example, a sample-size formula for 80% power for a two-sample independent-group t-test would be simply doubled for a four-group one-way analysis of variance (ANOVA). For an analysis of covariance (ANCOVA), the effect of the covariate would be ignored or assumed to increase power (an unwarranted assumption, in many cases). In a stratified design, the stratification would be ignored to use simple sample size methods. With the passage of time and the expansion of the statistical literature, more complicated sample size and power calculation methods have become known. In recent years, statistical software has become available that allows calculation for some complex designs.

Appropriate sample size and power methods for complicated analyses allow more precise accurate estimates of necessary sample size. Sometimes the proposed sample sizes are larger than those using simpler methods and sometimes they are smaller. The good news is that they allow for the most efficient research.

4. Effect Size

By definition, statistical power is the probability of rejecting H_0, the null hypothesis, in favor of H_1, the alternative hypothesis, when H_1 is true. It is intuitively obvious that if H_1 represents a large distance or difference from the situation under H_0, the sample size needed to detect a statistically significant difference at level α would be small. Conversely, if H_1 represents a small difference from H_0, the required sample size would be large.

For example, if the null hypothesis is that elephants weigh on average the same as mice, we would not need many specimens of elephants and mice to detect a statistically significant difference. Elephants and mice are far apart in terms of weight. However, to detect a difference in mean weight between two strains of mice, researchers would need a large sample of each type. The tiny difference in mean weight between two strains of mice would yield a very small effect size.

The term "effect size" has been used to signify the critical difference to be detected. In more formal usage, effect size defines a formula for the difference between H_0 and H_1 that is useful for sample size calculation. For example, for a two-group independent-sample t-test, the effect size is $(\mu_1 - \mu_2)/\sigma$, the difference in means in terms of the common standard deviation. This effect size, often called d, appears in power and sample size formulas. Obviously, different versions of sample size formulas for the same method may call for differing effect size formulas, but if differing effect sizes and their associated formulas are used correctly, they will yield identical sample sizes. Some common effect size formulas are given in **Table 1**.

From the formulas in **Table 1**, it is apparent that effect size, for each sample size calculation, represents a "distance" from the null value. The sample size for fixed power (say, 80%) will be larger for smaller effect sizes and smaller for larger effect sizes. This relationship is usually not linear, but of course depends on how effect size is represented in the sample size formula.

Figure 2 contains a graph of the relationship between effect size and sample size needed for 80% power in a one-sample t-test. As demonstrated in this plot, sample size varies extremely as the effect size increases from 0.1 to 1.0.

5. Steps in Sample Size Calculations

The following steps are a general guideline for sample size calculations.

1. Determine, in conjunction with other researchers, the research question. This may be very vague at first, but an attempt should be made to make it specific and quantitative. For example, the initial research question "Does hunger affect mood?" might be quantified to "How is blood glucose level related to mood, as measured by the Beck Depression Inventory?" *(16)*.
2. Reword the research question into a research hypothesis. Our glucose example hypothesis might be: "Glucose level is negatively correlated with Beck Depression Score."
3. Reverse the research hypothesis to form a null hypothesis. For example, in our glucose study:

$$H_0: \rho = 0$$

where ρ is the correlation between blood glucose and BDI score. Note that we have used a two-sided hypothesis, rather than the one-sided hypothesis suggested by the research hypothesis. Many researchers and journal editors prefer a two-sided hypothesis, which is considered to be more objective and conservative than a one-sided hypothesis.

Table 1
Some Effect Size Formulas

Statistical method	H_0	H_1	Effect size
One-group t-test or z-test	$\mu = 0$	$\mu = \mu_1 \neq 0$	μ_1/σ
Two-group t-test or z-test	$\mu_1 = \mu_2$	$\mu_1 \neq \mu_2$	$(\mu_1 - \mu_2)/\sigma$
Linear correlation	$\rho = 0$	$\rho = \rho_1 \neq 0$	ρ_1
K-group ANOVA	$\mu_1 = \cdots = \mu_k = \mu$	some difference among μ_i's	$\dfrac{\left(\sum\limits_1^k (\mu_i - \mu)^2/k\right)}{\sigma^2}$
Contrast in K-group ANOVA	$\sum\limits_1^k c_i \mu_i = 0$	$\sum\limits_1^k c_i \mu_i \neq 0$	$\dfrac{\left|\sum\limits_1^k c_i \mu_i\right|}{\sigma \sqrt{\sum\limits_1^k c_1^2}}$

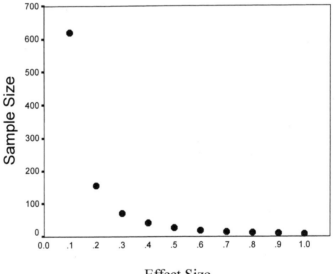

Fig. 2. Relationship between effect size and sample size needed for 80% power in a one-sample t-test.

4. Determine the significance level for the study. In most cases, $\alpha = 0.05$ will be the chosen level of significance, although some researchers choose $\alpha = 0.01$ or $\alpha = 0.02$ as a more conservative significance level.
5. Determine the statistical power to be used. As mentioned above, 80% or 90% is considered acceptable for most research studies.

6. Determine an important difference or distance for detection in the research study. A correlation of 0.3 or more, say, would be important to recognize in the glucose study.
7. Determine the statistical method to be used. Use the important difference proposed in **step 6**, and other estimates such as standard deviation, if necessary, to compute effect size. In the proposed glucose study, a correlation analysis is proposed, so the effect size is $\rho = 0.3$.
8. Calculate the sample size for the proposed study. The appropriate sample size for our correlation study is 85. Some methods of calculation are summarized in the paragraphs that follow.

At this point the initial sample size has been calculated. This critical number must be scrutinized.

5.1. Is the Calculated Sample Size Too Large to Be Practical?

Can we recruit this many subjects in a reasonable time frame? Can we afford to make this many expensive measurements?

If the sample size is unsatisfactorily large, an effort can be made to reduce it by alteration of the experiment plan. Researchers should investigate one or more of the following modifications, all of which are designed to decrease the required sample size:

1. Raise the level of significance (from, say, 0.01 to 0.05).
2. Decrease the desired power level (from, say, 90% to 80%).
3. Increase the effect size to be detected. A caveat here is that effect size should not be made *too* large to reduce the required sample size. For example, a correlation of 0.9 is almost never found in nature, although this effect size leads to a very small calculated sample size in correlation studies. An experiment designed to detect only unrealistically large differences will not be respected or valued.
4. If feasible, switch to a one-tailed test, which, in most cases, more powerful than its two-tailed counterpart. It is sometimes possible, on scientific grounds, to justify a one-sided test, although, as mentioned previously, a two-sided test is preferred by many.
5. Consider a redesign of the study, using more informative data, as discussed earlier. Reverting to original measurement data before cut-points are taken or substituting a research question involving a surrogate variable can lead to a substantial reduction in the required sample size, cutting the size in half or even reducing it 10-fold (*11*, pp. 275–276).
6. Consider a redesign of the experiment, using a different statistical method. For example, using repeated measures, crossover designs, or incorporating covariates for variance reduction can sometimes achieve significant savings in the required sample size (*17*).

All of these modifications must lead to a recalculation of the required sample size, which will, it is hoped, be reduced sufficiently for the experiment to be practicable.

Even if the initial stages of sample size determination yield sample sizes that are achievable and satisfactory to the researchers, it is often advisable to calculate sample sizes for several experiment designs. Each design, such as pre–post, crossover, repeated measures, and so on has advantages and disadvantages in research and sample sizes will vary. A careful, thoughtful sample size calculation will lead to the most efficient, informative research *(18)*.

If all attempts to address the research question with an experiment design with a practicable sample size fail, then the experiment should be abandoned. To proceed with an experiment whose sample size is insufficient to achieve a respectable level of statistical power is wasteful of resources and, in some instances, unethical.

5.2. Is the Calculated Sample Size Too Small?

This may seem an unlikely problem, but sometimes, particularly with interval measures, repeated measures, and correlation studies, the sample size from the initial calculation would not impress the peer reviewers or readers of the scientific literature. If the statistical method is an asymptotic one, such as a χ^2 test based on the normal approximation, it may be that the calculated small sample size would not be sufficient to provide accurate statistics and provide accurate approximations. If the sample size is too low, researchers should consider or discuss sample size recomputation using a smaller effect size, lower α-level, or higher statistical power. This might seem wasteful, but if correct research results cannot be published or respected, then the experiment has been a waste in terms of contributing to the body of knowledge.

Finally, it is advisable to boost the sample size slightly to allow for failed experiments or loss-to-follow-up. A rule of thumb is to add 10% for loss-to-follow-up *(19*, p. 1), but each laboratory or researcher should estimate this number from past experience in the research setting. Some sample size software (for example, Power and Precision, discussed in the following section) allows for built-in attrition adjustments.

6. Methods for Sample Size Calculations
6.1. Formulas

In some cases, a closed-form formula, usually involving the appropriate effect size, is available. Lachin *(20)* presents numerous formulas for sample size calculations. Biostatistics textbooks, such as those by Zar *(21)* and Dawson-Saunders and Trapp *(22),* as well as clinical trial books *(23,24)* contain sample size formulas. Friedman et al. *(24*, pp. 125–129) contains an extensive bibliography of articles with sample size methods.

Matrix-based formulas for power calculations for linear models appear in *(25)*. These formulas can be implemented using any computer-based matrix language that also incorporates functions for the noncentral F-distribution. SAS IML® is an example of such a language.

If one is faced with a power analysis for a new or complicated method, sometimes a search through the Current Index to Statistics *(26)* will yield a reference to an article detailing the appropriate power analysis. In some cases, power and sample size tables will be a part of the original article describing a new method.

6.2. Software

In recent years, several excellent computer packages have been developed for sample size and power calculations. The early versions of these programs provided power and sample size calculations mainly for simple models. The programs are under continuous development, improving user friendliness and the range of statistical methods covered. Up-to-date descriptions of the programs, as well as new programs not mentioned here, can be found on the World Wide Web. Some sample size and power programs and the software companies are:

nQuery *(27)*, an easy-to-use, versatile system	Statistical Solutions www.statsol.ie/nquery/nquery.htm
PASS (Power and Sample Size) *(28)*, user-friendly system with good graphics. Includes group sequential clinical trials	NCSS, Number Cruncher Statistical System www.ncss.com/
SamplePower *(29)*	SPSS, Inc. www.spss.com/spower/
Also marketed as Power and Precision *(30)* User-friendly, especially good for survival analysis, survival analysis, excellent graphics	Biostat www.powerandprecision.com

Other specialized programs that have some sample size calculation capabilities are:

EAST *(31)*, group sequential clinical trials	Cytel Software Corporation www.cytel.com
Egret Siz *(32)*, Cox regression and epidemiological models	Cytel Software Corporation www.cytel.com
Epi Info *(33)*, free downloadable epidemiological software	Centers for Disease Control and Prevention www.cic.gov/epiinfo

For the latest information regarding software capability and availability, visit the appropriate website or contact the individual or institution distributing the software.

Two excellent freeware programs for complicated linear models, multi-variate linear models including repeated measures, and other sample size problems are:

UnifyPow.sas	Ralph G. O'Brien, Cleveland Clinic Foundation
	www.bio.ri.cef.org/power.html
IML Power Program	Lynette L. Keyes and Keith E. Muller
	University of North Carolina
	ftp://ftp.uga.edu/pub/sas/contrib/cntb0014/

These programs, using SAS *(6)* macros, is are described in **ref. *34***.

Increasing software development is resulting in more programs for sample size determination. As of this writing, some sample size calculations are available on the World Wide Web. Using a search engine, it is possible to find websites that use Java Applets and other software to perform free sample-size calculations in real time.

A caveat with any computer program is to test it thoroughly if the methods and results upon which it is based have not been published. Testing can be accomplished by checking agreement with hand calculations, other software, or tables. All programs have disclaimers, absolving the authors and corporations involved in distribution of the software of any liability in case the results of the program are erroneous. The cost of an experiment with the wrong sample size will not be borne by software vendors or creators.

6.3. Tables

Books and journal articles with sample size and power tables are widely available. A list of books (and authors) with sample size tables appears below:

Statistical Power Analysis for the Behavioral Sciences (5), introduction and basic sample size calculations	Jacob Cohen
How Many Subjects? (9), basic sample size tables and explanations	Helena Chmura Kraemer and Sue Thiemann
CRC: Guide to Clinical Trials (19), concentrating on survival analysis	Jonathan J. Shuster

6.4. Nomograms

The following resources provide useful graphs for sample size determination:

Sample Size Choice (35), for ANOVA models	Robert E. Odeh and Martin Fox
"Nomograms for Calculating the Number of Patients Needed for a Clinical Trial with Survival as an Endpoint" *(36)*	David A. Schoenfeld and Jane R. Richter

6.5. Simulation

Simulation is the method of last resort for sample size calculation. This labor-intensive method involves writing computer programs for empirical, estimated power calculations. Simulation is sometimes necessary when the statistical method is very complex or when sample size formulas, software, tables, or nomograms are not yet available for a particular statistical method. Statistics is an academic discipline, and academic statisticians who develop new statistical methods will often publish articles and papers describing the methods as soon as they are developed and proven. Quick publication is desirable to make the method available for use as soon as possible and also to establish priority and to obtain academic credit for publication. A later article, by the same author or perhaps by another author, may detail the sample size calculation. In the absence of formulas or tables, power and sample size estimates may come from simulation, using a computer package or language. The steps for simulation are outlined as follows:

1. Decide on the null hypothesis, a statistical method, the levels for α and power, the alternative hypothesis reflecting the important difference of interest, and an initial sample size "guestimate" n.
2. Write a computer program to generate data sets of size n according to the distribution described by the alternative hypothesis. A rule of thumb here is to generate at least 1000 datasets of the appropriate sample size, although more would mean more accurate estimates, of course. With this simulation we are estimating a proportion, $1 - \beta$, the statistical power under H_1.
3. For each dataset, compute the statistical test of H_0. Keep a tally of the number of tests that reject and the number that fail to reject.
4. Calculate the percentage of the datasets that lead to rejection. This is the power estimate for sample size n for this distance from the null hypothesis. If power is too low, adjust n upward and repeat the simulation steps above.
5. Continue this process until a satisfactory sample size and power are achieved. The Statistical statistical packages SAS and S-Plus have been used for simulation studies, as well as programming languages such as FORTRAN and BASIC (*37*, p. 139).

7. Special Topics in Power and Sample Size Analysis

7.1. Achieved Power

Achieved power is a power estimate based on the results of a study. That is, the data in the study are used to generate an estimate of effect size and statistical power. Often, achieved power is not very informative. It is usually very high when the experiment is statistically significant. Indeed, some researchers opine that achieved power must equal 1 if a study is statistically significant, observing that power is the probability of rejecting the null hypothesis and the null hypothesis has already been rejected in the study (*38*).

However, an achieved power not equal to 1, but usually quite high, can be estimated using the data. SPSS *(7)* prints achieved power for some analyses on request.

If the null hypothesis is not rejected, the achieved power may be very low (approaching α) or attain some middle value. This power estimate is a rough evaluation of the adequacy of the sample size of the study if the difference or distance from H_0 that has been observed is approximately equal to a difference of research interest. That is, if the data show an important research effect, but it is not statistically significant, the achieved power will give an indication of how far from the desirable power of 80% or 90% this sample size is. If the power is low even when an important effect is demonstrated in the descriptive statistics calculated from the data, the sample size is probably far from adequate.

Achieved power in general is a biased, inflated power estimate. If a new experiment is to be planned using the present results, it is best to adjust the sample size estimates to compute obtain an unbiased estimate of statistical power and sample size for the new study *(39,* pp. 405–416).

7.2. Post Hoc Power

Post hoc power is the term used to describe power computed after the completion of an experiment. Researchers use some of the experiment results (say, observed standard deviations, correlations, or variances) to compute the power to detect an important, conjectured difference. Achieved power, discussed previously, is thus a special case of *post hoc* power, one in which the effect size is also estimated from the completed experiment.

When an experiment is not statistically significant and power and sample size were not calculated carefully before execution of the study, power can be estimated after the fact to determine if the experiment was sensitive enough to detect a difference of value to the researchers. Power computed *post hoc* is late, at best, and the situation is tantamount to scientific fraud if researchers report that their experiment was carefully planned when it was not. Some researchers deplore *post hoc* power, while some believe it has value to salvage nonsignificant results for reporting purposes if the sample size of an unplanned study happens by chance to have been adequate for a reasonable alternative hypothesis.

Another interpretation is that the nonsignificant experiment now functions as a pilot study, and the *post hoc* power calculation is the first step toward designing a new study of appropriate sample size *(9,* p. 25).

7.3. Pilot Studies

When estimates needed for power and sample size calculations are not available from the literature, from existing databases, or from previous experience in the laboratory or clinic, a pilot study may be in order to develop information needed for adequate estimation. The importance of pilot studies has been recognized in some research institutions by the provision of intramural grants to help new researchers and by special pilot study grant programs in United States federal granting agencies. For example, the National Institute of Mental Health has provided special grant support to develop effect sizes to improve future research.

These are the hallmarks of a good pilot study:

1. A pilot study should be small. A large pilot study is an oxymoron, as a large, expensive study should answer the research question, not just help to plan another study. Careful planning of a pilot study to provide a tight (narrow) confidence interval for a variance can lead to a larger sample size than would be feasible for the ultimate research question! Common sense must be the guide.
2. The pilot study should measure various outcomes, as researchers may find that a surrogate marker, described previously, may be more suitable and economical for their research question. Time to occurrence of an event of interest is sometimes estimated in pilot studies for planning purposes.
3. The pilot study should obtain all estimates needed for power and sample size analysis for any feasible research design. These estimates usually include sample means and standard deviations, but they also should include estimated correlations for repeated measurements if this might be a possible experimental design. Estimated correlation between measures or repeated measures has a large effect on sample size (*17*, p. 41).

7.4. Grant Applications

At the 1997 Joint Statistical Meetings of the American Statistical Association, Ralph O'Brien of the Cleveland Clinic led a group of discussants in a seminar entitled "Statistical Grantsmanship." Among the suggestions for applications seeking national funding were the following:

1. Use of sophisticated power and sample size calculations, appropriate for the ultimate planned statistical analysis.
2. Use of appendices in grant applications for long sample size formulas and theory.

Other suggestions for optimal power and sample size calculations for grant applications include:

1. Use of previous knowledge gained from intramural studies and pilot studies. Known colloquially as "sweat equity," the work invested in well-designed pilot studies can give grant applicants an advantage over others in the selection process.
2. Sensitivity analyses to demonstrate adequate statistical power if the assumptions in the primary power analysis are not met. For example, it would be valuable to demonstrate that the planned study affords reasonable power for a range of variance estimates. It is also wise to investigate several alternative designs to choose the best one (*18*, p. 1209).

3. A summary table at the end of the power analysis section of a grant application can summarize the value of n chosen, the power and sensitivity for primary hypotheses, and the estimated power for secondary hypotheses. Sometimes power for secondary hypotheses will not be high, owing to budget constraints.

7.5. Cost of Power Analysis

The length of this chapter, which merely outlines power and sample size considerations in planning a study, is an indication of the complexity of this subject. It is not unusual for the tasks of power and sample size estimation for complicated research plans or for designs using new statistical methods to require weeks of work for statisticians and researchers. It is important to allocate adequate time and resources for sample size calculation (*18*, p. 1224).

8. Examples
8.1. Segregation Analysis for Codominant Loci

This problem is suggested by an example in *Statistics in Human Genetics* by Pak Sham *(40)*. In this genetic problem, individuals with heterozygous inheritance at the locus are phenotypically different from homozygous individuals. Mendelian inheritance implies that the three phenotypes will appear as offspring from two heterozygous parents in the proportions 1/4, 1/2, 1/4, where the 1/4 fractions represent homozygous offspring and the 1/2 fraction represents heterozygous inheritance. The research question is, "Do the proportions in offspring of heterozygous parents differ from those expected in Mendelian inheritance?" The research hypothesis is, "Proportions of the three phenotypes differ from Mendelian inheritance." We have H_0: $p_1 = 1/4$, $p_2 = 1/2$, $p_3 = 1/4$. We wish to detect a "difference" reflected by $p_1 = 0.3$, $p_2 = 0.4$, $p_3 = 0.3$. We use nQuery *(27)* and select the program "Chi-square test of specified proportions in C categories." We choose $\alpha = 0.05$ and power = 0.80 and type these numbers into the input table along with the number of categories, $C = 3$. nQuery has an effect size calculator, so we enter the null hypothesis proportions and the alternative hypothesis proportions and obtain the effect size $\Delta^2 = 0.04$. The required sample size is $n = 241$.

8.2. Time to Onset of Disease in a Small Animal Model

In laboratory mice an allele has been identified that is associated with the onset of cancer. Researchers can induce a certain type of cancer by introduction of a virus. Animals will be tested for presence of the suspected allele. Then the virus will be injected into each mouse. Subjects will be observed daily for onset of cancer. We anticipate that the population of mice without the suspect allele will have a median age of onset of disease of 60 d. Mice with the allele are suspected to experience earlier onset, median = 45 d. Animals will

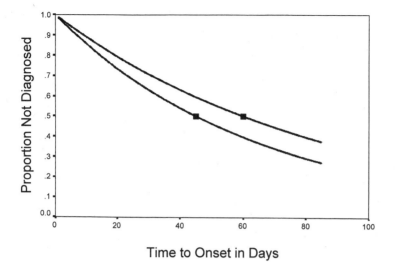

Time to Onset in Days

Fig. 3. Hypothesized survival curves and the median survival values for time until onset of disease.

be observed for 85 d. **Figure 3** presents the hypothesized survival curves and the median survival values.

The research question is, "Do subjects with the suspect allele tend to experience earlier onset of disease?" The research hypothesis is, "Animals with the suspect allele tend to experience earlier onset of disease." The null hypothesis is H_0: $S_1 = S_2$, where S_1 represents the survival curve for subjects with the allele and S_2 is the survival curve for subjects without the allele. We have chosen a two-sided hypothesis using the log-rank test.

We choose first to use nomograms for survival curves *(36)*. We compute $R =$ ratio of median survival times = 60/45 = 1.33. Using **Fig. 1**, p. 164, in **ref. *36***, for power = 0.8 and $\alpha = 0.05$, we draw lines according to the instructions for accrual = 0 (we will follow all animals for 85 d) and follow-up period = 1.6 ((85/.5 [(45 + 60)]/2]). We obtain 140 per group for $R = 1.5$, using the dashed line for two-sided hypotheses. Turning to **Fig. 3** of the paper, p. 166, in **ref. *36***, we perform an adjustment to obtain a sample size for $R = 1.33$: 300 per group. Thus, using the nomograms, we obtain $n = 600$.

Using the *CRC Handbook 19*, we must first compute the proportion without disease at 85 d. We compute λ, the hazard rate for exponential survival, according to the formula $\lambda = -(\ln 0.5)/(\text{median survival})$. We obtain $\lambda_1 = 0.0154$ and $\lambda_2 = 0.01155$. Using the formula $e^{-\lambda t}$ for the proportion not diagnosed at time t, we obtain 0.27 for the allele group and 0.37 for the controls (no suspect alleles) at $t = 85$. On p. 612 of **ref. *19*** for ALPHA = 0.025 (one-tailed, implying

0.05 ALPHA for two-tailed), PCONT = 0.25, DEL = 0.37 – 0.27 = 0.10, using FACT = 0.00 for up-front accrual, we find n = 579. For PCONT = 0.30, we find n = 665. Using linear interpolation, we have n = 613.

Finally, using the nQuery *(27)* program "Log-rank test of survival in two groups followed for fixed time, constant hazard ratio," and using the built-in parameter calculators to compute the λ's, we obtain 309 per group, or n = 618.

We have obtained similar sample sizes using three methods. All sample size methods hypothesized exponential survival. The sample size program nQuery was the easiest to use. An adjustment upward for loss-to-follow-up would be at the discretion of the researchers. Further refinements should be made if the proportions with and without the suspect allele are far from 0.50.

8.3. Validation of New Assay Method

Laboratory researchers wish to use a new, simpler assay method. They wish to establish that the new method affords a satisfactory level of accuracy. They decide to use Lin's concordance coefficient, which is preferable to correlation or *t*-test comparisons *(41)*. (*See* also Chapter 5 by Stephen W. Looney for further discussion of this issue.) Lin's original article defining the concordance coefficient was followed by another paper outlining sample size estimation and giving tables for sample size *(42)*. Using the guidelines in Lin's second paper **ref.** *42*, the researchers determine that they expect, under ideal conditions, that the new assay will explain 98% of the standard assay. They decide they can tolerate a 1% reduction in precision, a 12.5% location shift, and a 10% scale shift. Using the tables on p. 602 of *(42)*, they determine the minimum acceptable concordance, $\rho_{c,a}$, to be 0.972 and that a sample of 41 paired assays will be required for 95% power.

Acknowledgments

Thanks to Ralph O'Brien of the Cleveland Clinic for review of this chapter and excellent suggestions. Grateful appreciation also to Stephen Looney, my colleague at the University of Louisville, for his leadership and encouragement. Any remaining mistakes are my own.

References

1. Lehman, E. L. (1959) *Testing Statistical Hypotheses*. John Wiley & Sons, New York.
2. Lee, S. E. and Zelen, M. (2000) Clinical trials and sample size considerations: another perspective. *Statist. Sci.* **15,** 95–100.
3. Altman, D. G. (1994) The scandal of poor medical research. *Br. Med. J.* **308,** 283–284.
4. Freiman, J. A., Chalmers, T. C., Smith, H., Jr., and Kuebler, R. R. (1978) The importance of beta, the Type II error and sample size in the design and interpretation of the randomized control trial. *N. Engl. J. Med.* **20,** 690–694.

5. Cohen, J. (1987) *Statistical Power Analysis for the Behavioral Sciences*, 2nd ed. Lawrence Erlbaum, Hillsdale, NJ.
6. *SAS*, Statistical Analysis System, SAS Institute Inc., SAS Campus Drive, Cary, NC 27513.
7. *SPSS*, Statistical Package, for the Social Sciences, 233 South Wacker Drive, Chicago, IL 60606.
8. Beahrs, O. H., Henson, D. E., Hutter, R. V. P., Kennedy. (ed.) (1993) *Handbook for Staging of Cancer*. J. B. Lippincott, Philadelphia.
9. Kraemer, H. C. and Thiemann, S. (1987) *How Many Subjects? Statistical Power Analysis in Research*. SAGE, Newbury Park, CA, pp. 81–83.
10. Cohen, J. (1983) The cost of dichotomization. *Appl. Psychol. Meas.* 78, 240–253.
11. Goldsmith, L. J. (1995) Pros and cons of cutpoints, in *Proceedings of the Biometrics Section, Papers presented at the Annual Meeting of the American Statistical Association*, Orlando, FL, August 13–17, 1995, pp. 272–276.
12. Hughes, M D. (1999) The use and evaluation of suggogate endpoints in clinical trails, in *54th Deming Conference on Applied Statistics*, Atlantic City, NJ, December 8, 1999.
13. Lefkopoulou, M. and Zelen, M. (1995) Intermediate clinical events, surrogate markers and survival. *Lifetime Data Anal.* **1**, 73–86.
14. Topol, E. J., Califf, R. M., VandeWerf, F., Simoons, M., et al. (1997) Perspectives on large-scale cardiovascular clinical trials for the new millennium. *Circulation* **95**, 1072–1082.
15. Cohen, J. (1990) Things I have learned (so far). *Am. Psychol.* **45**, 1304.
16. Beck, A. T. and Steer, R. A. (1993) *Beck Depression Inventory: Manual*. Psychological Corporation, San Antonio, TX.
17. Venter, A. and Maxwell, S. E. (1999) Maximizing power in randomized designs when N is small, in *Statistical Strategies for Small Sample Research* (Hoyle, R. H., ed.) SAGE, Thousand Oaks, CA, pp. 31–58.
18. Muller, K. E., LaVange, L. M., Ramey, S. L., and Ramey, C. T. (1992) Power calculations for general linear multivariate models including repeated measures applications. *J. Am. Statist. Assoc.* **87**, 1209–1226.
19. Shuster, J. J. (1990) *Handbook of Sample Size Guidelines for Clinical Trials*. CRC Press, Boca Raton, FL.
20. Lachin, J. M. (1981) Introduction to sample size determination and power analysis for clinical trials. *Control Clin. Trials* 2, 93–113.
21. Zar, J. H. (1996) Biostatistical Analysis, Prentice Hall, Upper Saddle River, NJ.
22. Dawson-Saunders, B. and Trapp, R. G. (1994) *Basic and Clinical Biostatistics*. Appleton & Lange, Norwalk, CT.
23. Meinart, C. L. (1986) *Clinical Trials: Design, Conduct, and Analysis*. Oxford University Press, New York.
24. Friedman, L. M., Furberg, C., and Demets, D. L. (1996) *Fundamentals of Clinical Trials*, 3rd ed. Mosby, St. Louis, MO.
25. Graybill, F. A. (1976) *Theory and Application of the Linear Model*. Duxbury Press, North Scituate, MA.

26. Current Index to Statistics, Editors: Michael Wichura, Klaus Hinkelmann, http://www.statindex.org/

27. nQuery, Statistical Solutions, Janet D. Elashoff, Ph.D., Stonehill Corporate Center, Suite 104, 999 Broadway, Saugus, MA 01906.

28. PASS, Number Cruncher Statistical Systems, Jerry L. Hintze, PhD., 329 North 1000 East, Kaysville, UT 84037.

29. SamplePower, SPSS, Inc., 233 South Wacker Drive, 11th Floor, Chicago, IL 60606.

30. Power and Precision, Michael Borenstein, Director, Biostat, 14 North Dean Street, Englewood, NJ 07631.

31. EAST, Cytel Software Corp., 675 Massachusetts Ave., Cambridge, MA 02139.

32. Egret SIZ, Cytel Software Corp., 675 Massachusetts Ave., Cambridge, MA 02139.

33. Epi Info, The Division of Surveillance and Epidemiology, Epidemiology Program Office, Centers for Disease Control and Prevention (CDC), Atlanta, GA 30333.

34. O'Brien, R. G. and Muller, K. E. (1993) Unified power analysis for t-tests through multivariate hypotheses, in *Applied Analysis of Variance in the Behavioral Sciences* (Edwards, L. K., ed.). Marcel Dekker, New York, pp. 297–344.

35. Odeh, R. E. and Fox, M. (1991) *Sample Size Choice*, 2nd ed. Marcel Dekker, New York.

36. Schoenfeld, D. A. and Richter, J. R. (1982) Nomograms for calculating the number of patients needed for a clinical trial with survival as an endpoint. *Biometrics* **38,** 163–170.

37. Chow, S.-C. and Liu, J. (2000) *Design and Analysis Bioavailability and Bioequivalence Studies*, 2nd ed. Marcel Dekker, New York, p. 139.

38. Goodwin, S. N. and Berlin, J. A. (1994) The use of predicted confidence intervals when planning experiments and the misuse of power when interpreting results. *Ann. Intern. Med.* **121,** 202.

39. Winer, B. J. and Michels, K. M. (1991) *Statistical Principles in Experimental Design*, 3rd ed. McGraw-Hill. New York, pp. 405–416.

40. Sham, P. (1998) *Statistics in Human Genetics*. Arnold, London, p. 21.

41. Lin, L. I.-K. (1989) A concordance correlation coefficient to evaluate reproducibility. *Biometrics* **45,** 255–268.

42. Lin, L. I.-K. (1992) Assay validation using the concordance correlation coefficient. *Biometrics* **48,** 599–604.

7

Models for Determining Genetic Susceptibility and Predicting Outcome

Peter W. Jones, Richard C. Strange, Sud Ramachandran, and Anthony Fryer

1. Introduction

In this chapter, we focus on the study of associations between disease and inheritance of particular genetic variants, a field described as genetic epidemiology. Genetic variation based on polymorphism is common in human populations and appears to be a critical factor in determining susceptibility to disease. Polymorphism describes the presence of variant forms of genes (alleles) that are inhherited from parents. Individuals within a population may therefore inherit none (homozygous wild-type), one (heterozygote), or two (homozygous mutant) copies of the variant allele. These combinations are referred to as genotypes. Many types of allelic variation have been described, including deletions and insertions of DNA bases or even whole genes. Recently, genetic variation derived from single nucleotide polymorphisms (SNPs) — single base changes thought to occur every 500–1000 nucleotides — have attracted considerable interest in the context of disease susceptibility. For the purpose of this review, we will use data collected in our laboratories on ploymorphisms in members of the glutathione S-transferase (GST) supergene family of enzymes (*see* Hayes and Strange *[1]* for a recent review).

The number of genetic epidemiology association studies, and the complexity of statistical analysis of the resulting data, have developed exponentially in recent years. Many earlier studies were based on standard cross tabulations (chi-squared analyses) in small numbers of cases and controls, often with case groups being clinically heterogeneous. In many instances, even basic confounding factors such as age at presentation and gender were ignored, and most focused on a single gene *(2–6)*. Although this type of study still appears in journals, more recent work has identified the need for large, well-characterized

From: *Methods in Molecular Biology, vol. 184: Biostatistical Methods*
Edited by: S. W. Looney © Humana Press Inc., Totowa, NJ

patient cohorts with multivariate modeling to enable consideration of potential clinical heterogeneity within the total population *(7–11)*. This assessment of disease susceptibility continues to be plagued by problems of control selection and, in some cases, recruitment. Furthermore, much of the current success of this approach has highlighted the importance of examining outcome (e.g., disease severity, survival) rather than susceptibility alone (see the following paragraphs for examples).

2. General Modeling Concepts

In statistical terms, in any association study, we are given a series of candidate genes and possible confounders such as demographic (e.g., age, gender) and environmental (e.g., smoking habit, UV exposure) factors. Linear models are derived for determining whether (1) there are genetic components that affect susceptibility to disease and (2) within those who have the disease, outcome is also associated with the presence or absence of genetic risk factors (including gene–gene or gene–environment interactions) in the presence of the other nongenetic factors.

In multiple linear regression a dependent variable, outcome or response, y, is related to a set of independent variables, covariates or factors, $\mathbf{x} = (x_1, x_2, \ldots, x_p)$, where these are considered fixed, through a series of unknown parameters $\beta_1, \beta_2, \ldots, \beta_p$, such that the probability distribution of the random variable y has mean given by $\theta = \alpha + \Sigma \beta_i x_i$ and usually constant variance. A further assumption that is commonly made is that the y's are normally distributed. Some of the independent variables could be squares or cross-products of a smaller set of variables, so, for example, a linear regression can be constructed in which there is only one independent variable, x, and $\theta = \alpha + \beta_1 x + \beta_2 x^2 + \beta_3 x^3$.

McCullagh and Nelder *(12)* have extended this regression model to the generalized linear model (GLM) to deal with situations where the observations, y, are not continuous or normal. However, these models do have a common assumption that their mean, θ, is linear in the unknown parameters. These ideas are exploited in this chapter to develop three linear models, one for measuring susceptibility effects and the others for exploring disease outcome. We have applied these modeling concepts to specific examples taken from a study of the effects of GST polymorphisms on skin cancer risk in renal transplant patients (**Table 1**). All models produce estimates of genetic effects that may be interpreted as assessments of risk associated with genotypes but that are adjusted for other baseline characteristics.

3. Susceptibility

3.1. Clinical Considerations

Traditionally, case-control studies have compared the proportions of genotypes in a sample of cases with those in a sample of controls (often laboratory

Table 1
Study Examining the Association of Two Genes,
GSTM1 **and** ***GSTP1*,** **with Risk of Cutaneous Squamous**
Cell Carcinoma (SCC) in Renal Transplant Patients

(a) Risk of any SCC (logistic regression)
    logistic sccyn sex agetx M1n

Variable	OR	95% CI	p value
sex	3.8	1.0, 15.1	0.058
agetx	1.063	1.021, 1.106	0.003
M1n	3.2	1.1, 9.5	0.040

(b) Numbers of SCC lesions (negative binomial regression)
    nbreg scc_no sex agetx P1AA, exposure (folup) irr

Variable	RR	95% CI	p value
sex	16.6	2.7, 101.7	0.002
agetx	1.081	1.035, 1.129	< 0.001
P1AA	6.9	2.1, 22.8	0.002

Likelihood ratio test of $\gamma = 0$ (overdispersion test): $\chi^2_1 = 86.39$, $p < 0.001$

(c) Time from transplantation to development of the first SCC
    (Cox's proportional hazards regression)
    cox time sex agetx scor3 M1n M1nscor3, dead(censor)

Variable	HR	95% CI	p value
sex	3.7	0.9, 15.3	0.066
agetx	1.140	1.071, 1.213	< 0.001
scor3	0.17	0.03, .99	0.048
M1n	0.1	0.02, 0.9	0.041
M1nscor3	42.6	3.8, 479.7	0.002

Key

logistic	Stata command for logistic regression
sccyn	1 = SCC present, 0 = SCC absent
sex	1 = male, 0 = female
agetx	age at transplantation in years
M1n	1 = null genotype at *GSTM1* gene, 0 = other *GSTM1* genotypes
nbreg	Stata command for negative binomial regression
scc_no	number of SCC tumors
P1AA	1 = *AA* genotype at *GSTP1* gene 0 = other *GSTP1* genotypes
exposure	Stata command allowing normalization for exposure
folup	length of follow-up
irr	Stata command indicating display of incidence rate ratio
cox	Stata command for Cox's proportional hazards regression
time	time between transplantation and development of SCC (or last follow-up)
scor3	1 = sunbathing score ≥ 3, 0 = sunbathing score < 3
M1nscor3	1 = sunbathing score ≥ 3 and null genotype at *GSTM1* 0 = other genotype/sunbathing score combinations
dead	Stata command that specifies censor variable
censor	censor variable; 1 = SCC, 0 = no SCC

volunteers) by using chi-squared (χ^2) tests. For example, in an early study, we examined the frequency of genotypes at the *GSTM1* locus in hospital controls and patients with skin cancer using this approach *(13)*. These data showed that the proportion of patients with the *GSTM1* null genotype was significantly higher in patients with multiple skin cancers of different histological types (32/45, 71%) than in controls (79/153, 52%, $p = 0.033$). From these data, it is possible to obtain an assessment of the degree of risk imparted by this genotype by calculation of the odds ratio (OR = 2.3) and its corresponding confidence interval (95% CI = 1.1–4.8). These data can then also be used to assess the relative importance of the gene in determining susceptibility by attributable risk calculations *(14)*.

The importance of selection of suitable controls is demonstrated by studies of the sulfotransferase, *SULT1A1*, gene. A polymorphism in this gene has been shown to exhibit differences in genotype frequencies with increasing age in healthy controls *(15)*. It is therefore critical when studying associations between polymorphisms that cases be age-matched (preferably one-to-one, but at least overall) to controls, thereby necessitating more advanced statistical approaches (*see* **Subheading 6.**).

More recently, case-control data have been handled using more sophisticated statistical approaches, including examining the interactions between genetic and environmental factors (discussed in detail in *[16]*), with correction for confounding using multivariate models. For example, GST have traditionally been viewed as carcinogen detoxifying enzymes and several studies have examined the effect of GST genotype on lung cancer risk in smokers compared with nonsmokers *(7,17)*. In addition, the use of multivariate models to correct for confounding factors can provide useful information on the potential mechanism of the observed genetic association *(11,18)*. This is illustrated in **Table 1** (a) in which, using logistic regression analysis, the association of GSTM1 genotype with squamous cell carcinoma (SCC) risk in renal transplant recipients is corrected for the potential confounding effect of age at transplantation and gender, both of which are known SCC risk factors *(19)*.

3.2. Modeling Aspects

Models for susceptibility are developed using data from case-control (retrospective) studies (*see* Schlesselman *[20]* and Breslow and Day *[21]* for a comprehensive discussion of this type of study). In these models, the dependent variable is binary, with $y = 0$ representing the controls and $y = 1$ representing the cases in a logistic model.

This model makes the assumption that the probability of being a case, p, is related to **x** through the log odds or logit transformation given by

$$\text{logit}(p) = \log[p/(1-p)] = \theta = \alpha + \Sigma \beta_i x_i$$

Suppose that we wish to estimate the effect on y of a change in a binary variable, x_k from 0 to 1. (In the context of genetic susceptibility, x_k represents the presence or absence of a genotype.) If all other x's remain the same, the parameter β_k is:

$$\beta_k = \log \text{ odds } (x_k = 1) - \log \text{ odds } (x_k = 0) = \log [\text{ odds } (x_k = 1)/\text{odds } (x_k = 0)]$$

or the "log odds ratio." Rewriting, the odds ratio, which is a measure of the relative risk of being a case when an individual possesses a genotype, is e^{β_k}. The parameter β_k is estimated in the presence of all the other x's and may therefore be interpreted as a relative risk corrected for all other x's, unlike the uncorrected values in the previous subheading. These x's will have an effect on the estimate of risk and will be dealt with by the estimation procedure but may call into question the make up of the control group. In most disease association studies, a younger control group could become cases over time. Matching controls on key independent variables one-to-one with the cases is a possibility; this would lead to estimation using conditional as opposed to the unconditional logistic regression described above.

There are problems with applying this logistic model and indeed with the interpretation of simple odds ratios from 2 × 2 tables in case-control studies. As pointed out in Breslow and Day (*21*, Section 6.3.), the subjects are selected on the basis of whether they are cases or not, hence y cannot be regarded as a random variable while the x's, determined retrospectively, are random variables. This is the opposite of the usual situation. However, Breslow and Day show that, in most cases, the method of estimation of the logistic parameters which would be applied under the assumption that y is random or the data are generated from a cohort, or prospective, study gives similar numerical values of the parameter estimates and relative risks.

Table 1 illustrates the estimation of odds ratios (OR) for the GSTM1 null genotype using logistic regression modeling. The value of the odds ratio provides a quantitative estimate of the relative impact of the genetic effect on disease susceptibility. Many case-control studies on single genes in complex disorders such as cancer have often generated only modest odds ratios (e.g., 2.0–3.0). This suggests that other parameters must be taken into consideration. These include (1) the interaction of multiple genetic factors with each other and with environmental factors such as smoking, diet, or UV exposure and (2) the effect of genetic factors in genetically high risk subgroups diluted by a large number of cases with low genetic risk.

4. Outcome

4.1. Clinical Considerations

Studies in the literature on the same gene in the same disease group often produce conflicting data. For example, several groups have examined the importance of polymorphism in the GSTT1 gene in mediating susceptibility to colon cancer. Chenevix-Trench et al. *(22)* found that the frequency of the GSTT1 null genotype was significantly increased in patients diagnosed before 70 yr of age, while Deakin et al. *(23)* showed the frequency of this genotype was increased in cases compared with controls, although no age effect was observed. In contrast, Gerdig et al. *(24)* and Katoh and Bell *(25)* failed to show any significant association. These studies have highlighted the role of clinical heterogeneity, as recent data from our laboratory suggest that these discrepancies may result from differences in the proportion of patients with advanced disease in the different study populations. Thus, it is possible that, for example, GSTT1 null genotype is associated with poor outcome in these patients and, consequently, its frequency is increased in patients with advanced tumors (as reflected in Dukes' stage, tumor site, or age at onset). Therefore, the relative proportions of patients with advanced vs early disease in a sample of cases may determine whether the frequency of the polymorphism differs significantly between the cases and controls.

These observations have led to many studies examining the effect of so-called "modifier genes" on disease outcome. In general, these studies have met with significantly greater success that those examining susceptibility, with larger odds ratios (or other effect sizes). For example, **Table 1** (b),(c) show the association of *GST* genotype with both number of tumors and time between transplantation and appearance of the first SCC in renal transplant recipients. Number of tumors was examined using negative binomial regression, which, using the Stata statistical software package, allows simultaneous normalization for follow-up time (or degree of exposure) as well as correction for potential confounding factors. In this case, *GSTP1 AA* genotype has a rate ratio *(RR)* of 6.9. This represents a very large increase in risk, which may have clinical, as well as statistical, significance.

In the data described in **Table 1** (c), time between transplantation and appearance of the first SCC in renal transplant recipients was modeled using Cox's proportional hazards regression, an approach often used in survival analysis *(7,26)*. This example illustrates the examination of a gene–environment interaction between *GSTM1* null genotype and a high sunbathing score. This suggests that individuals who have high UV exposure and are *GSTM1* null genotype demonstrate a markedly reduced time from transplantation to development of their first SCC. Interestingly, when *GSTM1* null and sunbathing score were considered individually, neither was significant. This illustrates that an interaction cannot be predicted from the individual effects, a phenomenon referred to as *epistasis (27)*.

Thus, examination of associations between polymorphic variants and outcome may require the application of other types of statistical modeling. Indeed, we have used several approaches for the analysis of trends (ordered logistic regression) *(10)*, count data (Poisson regression and negative binomial regression) *(28,29)*, and time until an event (Cox's proportional hazards regression) *(18,30)*. We discuss some of the modeling aspects in more detail in the following subheading.

4.2. Modeling Aspects

In this subheading, we focus on the two modeling approaches illustrated in **Table 1** (b) and (c) (i.e., the accrual rate of tumors over time and the time to the next tumor) in patients with differing follow-up times. The methods may be applied to any situation where the dependent variable is a count or the time between two events.

When the data are in the form of counts and it is required to model associations with a set of independent predictors, there are a number of alternatives. Linear regression could be used with the square root of the counts as the dependent variable, or either Poisson or negative binomial regression could be used, where the means of the probability distributions are again a linear function of unknown parameters, some of which may be interpreted as measuring risk; we will concentrate on the latter two methods (*see [31]* for an application of these models to tumor counts). In each of these models, the time of exposure or follow-up will be a determinant of the eventual count and will have to be controlled for in the final estimation procedure.

In Poisson regression, the rate at which incidents (tumors) occur, usually termed the incidence rate, is assumed to be

$$\lambda = e^{\theta} = e^{\alpha + S\beta_i x}$$

for an individual with independent variable vector x. It follows that the mean of the number of incidents will be λT, where T is the exposure of the individual. To measure the effect of x_k changing from 0 to 1 (or a unit increase in x_k) where all other x's remain the same, the incidence rate ratio *(irr)* may be calculated as (incidence rate when $x_k = 1$)/(incidence rate when $x_k = 0$). It follows that $irr = e^{\beta_k}$. This may now be used as a way of evaluating the risk, in the sense of the effect of a change in the value of x_k producing a decrease or increase in the mean number of incidents, since for the same exposure, T, the *irr* is the ratio of the (corrected) number of incidents in the two groups (those with $x_k = 1$ and those with $x_k = 0$). This may be applied directly to the skin cancer study in **Table 1** to evaluate the effect of the presence of a genotype, in the presence of other covariates, on tumor incidence.

The Poisson distribution has the property that the mean and variance are equal and so if samples of the number of incidents in a study produce means and variances which differ, then this suggests that the assumption of a Poisson model for the counts is not tenable. More formally, a χ^2 goodness of fit test may be used to determine whether the model is suitable. In the situation where the variance is larger that the mean (or the process is exhibiting extra Poisson variation or overdispersion; underdispersion is rare), the negative binomial regression model can be used. This is a modification of the Poisson where the mean is multiplied by a random variable, z, whose distribution depends on a single overdispersion parameter γ (*see* **Table 1** [b]). This leads to a mean of $z\lambda T$.

If the measure of severity is the time between presentation with the first tumor and occurrence of the next tumor, then methods in survival analysis are appropriate. These methods have been developed to use data on all individuals, including those where the second event has not yet occurred but where a follow-up time is available. The parameter of interest here is the hazard rate, $\lambda(t)$, which measures the risk of an incident occurring at a particular point in time; another interpretation is that it is the instantaneous incident rate. In many practical situations this rate will vary with time. A full exposition of survival analysis may be found in Parmar and Machin *(32)*, which includes many clinical examples. They give an example of a hazard function and its relationship with time for measuring the risk of infant mortality, which is known to be highest just after birth but thereafter declines rapidly.

Cox's proportional hazards regression (Parmar and Machin *[32]*, Cox *[33]*), may be used to assess the impact of genotypes on time to a further event, adjusted for other x's. It is assumed that the hazard takes the form

$$\lambda(t) = \lambda_0(t)\exp(\Sigma\beta_i\, x_i),$$

where $\lambda_0(t)$ is interpreted as an underlying hazard for all individuals that is adjusted by the x's to give the hazard for a individual. In the linear form employed above $\alpha = \ln\lambda_0(t)$. The risk associated with a change of x_k from 0 to 1, with all other x's unchanged, is measured by the hazard ratio (HR) given by (hazard when $x_k = 1$)/(hazard when $x_k = 0$), which is easily seen to be e^{β_k}. In **Table 1** (c), we present HRs for two main effects and an interaction. The main assumption behind the Cox model is that the hazards are proportional; this is usually tested by a visual inspection of graphs of the estimated hazard functions in the two groups based on the values of x_k. If these are approximately parallel, then it suggests that the assumption is reasonable.

Because all models in this section depend on a linear function of covariates then it is possible to use the estimates of these functions to derive prognostic indices or simple scoring models for outcome. Christiansen *(34)* gives a detailed description of using Cox's regression to derive a prognostic index (PI) for an individual with a given set of x's.

5. Software

Most commercial statistics packages include routines for the analysis of linear, logistic, and Cox's proportional hazards regression models. However, few offer the possibility of fitting Poisson and negative binomial regression models. Stata (Stata Corporation, College Station, TX) offers a programmable package that includes these routines; all the analyses presented here were performed using this package and **Table 1** gives the command code necessary to generate the results.

6. Extensions to the Basic Modeling Approach

Most of the examples given in this chapter and the Stata code in **Table 1** refer to only a single genotype. It is straightforward to extend this to large multivariate models with other genotype main effects and interactions between genotypes. Furthermore, most packages will allow the user to specify stepwise selection of predictors. This may be used in cases where it is useful to allow all predictors to compete to enable a best subset of them to be chosen.

This could be especially useful in obtaining an estimate of the relative importance of genetic *vis à vis* environmental factors. However, care should be exercised in using these routines, especially where there are missing values in the data, as inferences could be based on only a small percentage of the observations. In this case, we suggest refitting the model with all the data using the reduced set of predictors.

We have briefly discussed the difficulty of obtaining suitable controls and the need to match or control for key variables when looking at susceptibility. However, this does not address the problem of population stratification in which the genetic background of the cases may be different from that of the controls. One method that is used in genetic studies that reduces the impact of this effect is to use a family-based study design such as that employed in transmission disequilibrium testing (TDT; *see* **[ref. 35**, Section 4.7] for more details). This approach compares the proportions of transmitted and nontransmitted alleles in parents and affected offspring, thereby correcting for the potential effect of non-disease-associated genetic differences between cases and controls including ethnic, geographical, and in some cases exposure differences.

It is worth noting that the studies described in this chapter are association studies. Thus, it is possible that a genotype demonstrates a strong association with risk without directly affecting the disease process at all. Indeed, the genetic marker may simply be coinherited with a neighboring gene that is critical to the pathological process.

References

1. Hayes, J. D. and Strange, R. C. (2000) Glutathione S-transferase polymorphisms and their biological consequences. *Pharmacology* **61**, 154–166.
2. Davies, M. H., Elias, E., Acharya, S., Cotton, W., Faulder, G. C., Fryer, A. A., and Strange, R. C. (1993) GSTM1 null polymorphism at the glutathione S-transferase M1 locus: phenotype and genotype studies in patients with primary biliary cirrhosis. *Gut* **34**, 549–553.
3. Yu, M-W., Gladek-Yarborough, A., Chiamprasert, S., Santella, R. M., Liaw, Y-F., and Chen, C-J. (1995) Cytochrome P450 2E1 and glutathione S-transferase M1 polymorphisms and susceptibility to hepatocellular carcinoma. *Gastroenterology* **109**, 1266–1273.
4. Katoh, T., Inatomi, H., Nagaoka, A., and Sugita, A. (1995) Cytochrome P4501A1 gene polymorphism and homozygous deletion of the glutathione S-transferase M1 gene in urothelial cancer patients. *Carcinogenesis* **16**, 655–657.
5. El-Zein, R. A., Zwischenberger, J. B., Abdel-Rahman, S. Z., Sankar, A. B., and Au, W. W. (1997). Polymorphism of metabolising genes and lung cancer histology: prevalence of CYP2E1 in adenocarcinoma. *Cancer Lett.* **112**, 71–78.
6. Nazar-Stewart, V., Motulsky, A. G., Eaton, D. L., White, E., Hornung, S. K., Leng, Z. T., et al. (1993) The glutathione s-transferase-mu polymorphism as a marker for susceptibility to lung-carcinoma. *Cancer Res.* **53**, 2313–2318.
7. Goto, I., Yoneda, S., Yamamoto, M., and Kawajiri, K. (1996) Prognostic significance of germ line polymorphisms of the *CYP1A1* and glutathione S-transferase genes in patients with non-small cell lung cancer. *Cancer Res.* **56**, 3725–3730.
8. Brockmoller, J., Cascorbi, I., Kerb, R., and Roots, I. (1996) Combined analysis of inherited polymorphisms in arylamine N-acetyltransferase 2, glutathione S-transferases M1 and T1, microsomal epoxide hydrolase, and cytochrome P450 enzymes as modulators of bladder cancer risk. *Cancer Res.* **56**, 3915–3925.
9. Probst-Hensch, N. M., Haile, R. W., Ingles, S. A., Longnecker, M. P., Han, C. Y., Lin, B. K., et al. (1995). Acetylation polymorphism and prevalence of colorectal adenomas. *Cancer Res.* **55**, 2017–2020.
10. Fryer, A. A., Bianco, A., Hepple, M., Alldersea, J., Jones, P. W., Strange, R. C., and Spiteri, M. A. (2000) Polymorphism at the glutathione S-transferase, GSTP1, locus is associated with atopy/airway responsiveness. *Am. J. Respir. Crit. Care Med.* **161**, 1437–1442.
11. Clairmont, A., Sies, H., Ramachandran, S., Lear, J. T., Smith, A. G., Bowers, B., et al. (1999) Association of NAD(P)H:quinone oxidoreductase NQO1 null with numbers of basal cell carcinomas: Use of a multivariate model to rank the relative importance of this polymorphism and those at other relevant loci. *Carcinogenesis* **20**, 1235–1240.
12. McCullagh, P. and Nelder, J.A. (1989) *Generalised Linear Models*, 2nd ed. Chapman and Hall, London.

13. Heagerty, A. H., Fitzgerald, D., Smith, A., Bowers, B., Jones, P., Fryer, A. A., et al. (1994) Glutathione S-transferase GSTM1 phenotypes and protection against cutaneous tumors. *Lancet* **343**, 266–268.

14. Hutchinson, P. E., Osborne, J. E., Lear, J. T., Smith, A. G., Bowers, P. W., Jones, P. W., et al. (2000) Vitamin D receptor polymorphisms are associated with altered prognosis in patients with malignant melanoma. *Clin. Cancer Res.*, **6**, 498–504.

15. Coughtrie, M. W. H., Gilissen, R. A. H. J., Shek, B., Strange, R. C., Fryer, A. A., Jones, P. W., and Bamber, D. E. (1999) The phenol sulfotransferase *sult1a1* polymorphism: molecular diagnosis and allele frequencies in Caucasian and African populations. *Biochem. J.* **337**, 45–49.

16. Fryer, A. A. and Jones, P. W. (1999) Interactions between detoxifying enzyme polymorphisms and susceptibility to cancer, in: *Metabolic Polymorphisms and Cancer, vol. 148* (Boffetta, P., Caporaso, N., Cuzick, J., Lang, M., and Vineis, P. eds.) IARC, Lyon, pp. 303–322.

17. Kihara, M., Kihara, M., and Noda, K. (1995). Risk of smoking for squamous and small cell carcinomas of the lung modulated by combinations of CYP1A1 and GSTM1 gene polymorphisms in a Japanese population. *Carcinogenesis* **16**, 2331–2336.

18. Howells, R. E. J., Redman, C. W. E., Dhar, K. K., Sarhanis, P., Musgrove, C., Jones, P. W., et al. (1998) Association of glutathione S-transferase GSTM1 and GSTT1 null genotypes with clinical outcome in epithelial ovarian cancer. *Clin. Cancer Res.* **4**, 2439–2445.

19. Ramsay, H. M., Harden, P. N., Reece, S., Smith, A. G., Jones, P. W., Strange, R. C., and Fryer, A. A. (2000) Polymorphisms in glutathione S-transferases are associated with altered risk of non-melanoma skin cancer in renal transplant recipients: a preliminary analysis. *J. Invest. Dermatol.*, in press.

20. Schlesselman, J. J. (1982) *Case-Control Studies.* Oxford University Press, New York.

21. Breslow, N. E. and Day, N. E. (1980) *Statistical Methods in Cancer Research,* vol. 1: *Case Control Studies.* IARC, Lyon.

22. Chenevix-Trench, G., Young, J., Coggan, M., and Board, P. (1995) Glutathione S-transferase M1 and T1 polymorphisms: susceptibility to colon cancer and age of onset. *Carcinogenesis* **16**, 1655–1657.

23. Deakin, M., Elder, J., Hendrickse, C., Peckham, D., Baldwin, D., Pantin, C., et al. (1996) Glutathione S-transferase GSTT1 genotypes and susceptibility to cancer: studies of interactions with GSTM1 in lung, oral, gastric and colorectal cancer. *Carcinogenesis* **17**, 881–884.

24. Gertig, D. M., Stampfer, M., Haiman, C., Hennekens, C. H., Kelsey, K., and Hunter, D. J. (1998) Glutathione S-transferase GSTM1 and GSTT1 polymorphisms and colorectal cancer risk: a prospective study. *Cancer Epidemiol. Biomarkers Prev.* **7**, 1001–1005.

25. Katoh, T. and Bell, D. A. (1996) Glutathione S-transferase M1 and T1 genetic polymorphism and susceptibility to gastric and colorectal adenocarcinoma. *Proc. Am. Assoc. Cancer Res.* **37**, 257–258.

26. Ramsay, H. M., Fryer, A. A., Smith, A. G., and Harden, P. N. (2000) Clinical risk factors associated with non-melanoma skin cancer in renal transplant recipients. *Am. J. Kid. Dis.* **36,** 167–176.

27. Frankel, W. N. and Schork, N. J. (1996) Who's afraid of epistasis? *Nat. Genet.* **14,** 371–373.

28. Yengi, L., Inskip, A., Gilford, J., Alldersea, J., Bailey, L., Smith, A., et al. (1996) Polymorphism at the glutathione S-transferase, GSTM3 locus: interactions with cytochrome P450 and glutathione S-transferase genotypes as risk factors for multiple cutaneous basal cell carcinoma. *Cancer Res.* **56,** 1974–1977.

29. Ramachandran, S., Lear, J. T., Ramsay, H., Smith, A. G., Bowers, B., Hutchinson, P. E., et al. (1999) Presentation with multiple cutaneous basal cell carcinomas: association of glutathione S-transferase and cytochrome P450 genotypes with clinical phenotype. *Cancer Epidemiol. Biomarkers Prev.* **8,** 61–67.

30. Lear, J. T., Smith, A., Heagerty, A. H. M., Bowers, B., Jones, P. W., Gilford, J., et al. (1997) Truncal site and detoxifying enzyme polymorphisms significantly reduce time to presentation of next cutaneous basal cell carcinoma. *Carcinogenesis* **18,** 1499–1503.

31. Drinkwater, N. R. and Klotz, J. H. (1981) Statistical methods for the analysis of tumor multiplicity data. *Cancer Res.* **41,** 113–119.

32. Parmar, M. K. B. and Machin, D. (1995) *Survival Analysis: A Practical Approach.* John Wiley and Sons, Chichester.

33. Cox, D. R. (1970) Regression models and life tables (with discussion). *J. Roy. Statist. Soc. Ser. B* **34,** 187–220.

34. Christiansen, E. (1987) Multivariate survival analysis using Cox's regression model. *Hepatology* **7,** 1346–1358.

35. Sham, P. (1998) *Statistics in Human Genetics.* Arnold Press, London.

8

Multiple Tests for Genetic Effects in Association Studies

Peter H. Westfall, Dmitri V. Zaykin, and S. Stanley Young

1. Introduction

Many common human diseases have a genetic component as measured by familial studies. Metabolic disorders such as diabetes, cardiovascular diseases such as high blood pressure, psychiatric disorders such as schizophrenia, and neurodegenerative diseases such as Alzheimer's disease all are thought to have a hereditary component. In some diseases the genetic control is through a single gene, while in others, multiple genes interact in complex ways with environmental factors to produce the disease *(1–5)*.

Data are and will become increasingly available to attempt to link genes to disease phenotype(s). Linkage studies, although powerful for screening relatively large chromosomal regions, lack needed precision because of the constraints imposed by the number of recombination events during generations contained in the pedigree *(6)*. Recently, researchers have attempted to develop techniques that exploit possibilities of fine mapping due to linkage disequilibrium between genetic markers and disease genes. Typing single nucleotide polymorphism markers (SNPs) inside of candidate regions provides a potential means for such analysis *(7)*; however, the problem remains in that the complex diseases are very likely to have multiple etiologies. Consider control of essential hypertension. It has a measured heritability of 35–45%, yet the identification of specific genes remains unclear. Many candidate genes for essential hypertension have been identified and, in a particular individual, a combination of some few of these genes might lead to disease.

There is a need for a statistical strategy to analyze these complex experiments, given the multiple testing implied by multiple candidate genes and the risk of false associations. In this chapter we discuss primarily methods for controlling

From: *Methods in Molecular Biology, vol. 184: Biostatistical Methods*
Edited by: S. W. Looney © Humana Press Inc., Totowa, NJ

familywise error rate (FWE) with multiple genetic tests, identifying single and epistatic effects, and discuss readily available software (PROC MULTTEST of SAS/STAT®) for this purpose. The benefit of the method is sound inference in the evaluation of case-control genotype–phenotype association studies.

2. Multiple Testing Principles for Disease–Genotype Association

Our focus is primarily on multiple contingency table-type tests described in, for example, Sasieni *(8)*, and extensions thereof. In the simplest analysis, subjects are cross-classified in a 2×2 table, according to disease status (case or control) and presence or absence of a particular allele at a given locus.

As an initial screening procedure, one may perform a test for each genetic locus in a genome scan or dense SNP map. Such tests are associational only, and further study is needed to establish causation; however, they can be very useful to identify candidate genes. Follow-up analyses can proceed using, for example, linkage analysis or haplotype-level tests *(9)*.

When these tests are performed separately over multiple loci, there can be hundreds, even thousands, of tests, and false-positives are expected, as discussed throughout the statistics and genetics literature (e.g., *10–12*). Various methods have been proposed to control this risk in genetic studies, such as FWE-controlling methods *(12)*; informal, global-based testing methods *(13)*; and false discovery rate (FDR) controlling methods *(14)*.

We suggest controlling FWE and justify it in two ways. First, control of FWE has a simple operational interpretation: If the FWE is set at 10% (say), then we expect that in only one out of every 10 studies will one or more false significant results be claimed. Therefore, the analyst may gamble upon the occurrence that the given study was not one of those 10%, and claim that all identified associations are real and repeatable. The FDR controlling procedure of Benjamini and Hochberg *(15)*, described in Weller et al. *(14)* for genetic QTL analysis, while more powerful than FWE for gene finding, does not allow such a clear operational definition. In a given study, the number of erroneous significances is a random variable, and therefore somewhat unpredictable. Furthermore, while FDR-controlling methods allow only an average of $100\alpha\%$ of the claimed significances to be in error, the false discovery rate can be substantially larger than that in studies where one or more genes have been declared significant *(16)*. Thus, although FDR-controlling methods are indeed more powerful, their operational interpretation is not as useful as that of FWE-controlling methods.

Second, advances in modern computing have made the powerful FWE-controlling "closed testing" methods accessible for the analysis of genetic tests. In particular, these methods can accommodate discreteness and genetic correlation structures (including linkage) to improve power. In our examples we will incorporate such features through exact testing methods.

For readers unfamiliar with multiple testing methods, closed testing, and/or PROC MULTTEST, it may be helpful to read Westfall and Wolfinger's (2000) article "Closed Multiple Testing Procedures and PROC MULTTEST," available on the SAS® website (http://www.sas.com/service/library/periodicals/obs/obswww23/). The remainder of this chapter is a condensed summary of material therein, with special emphasis on genetics applications.

2.1. The Closure Principle

FWE-controlling methods can be made less conservative and more powerful by using the closure principle of Marcus et al. *(17)*. The procedure is as follows: one considers all possible combination hypotheses obtained via intersection of the set of base hypotheses of interest. If the base hypothesis, and all intersections that contain it as a subcomponent, are all rejected by an appropriate α-level test (we will use exact tests here), then the closure principle allows that the given hypothesis can be rejected, at FWE level α. Thus, if there are k base tests, there are 2^k-1 tests to consider. For small studies, this procedure is ideal; however, for typical genotype/phenotype association studies where thousands of genotypes are considered, the number of intersection subsets to evaluate seems astronomical, and uncomputable even by current standards. However, there are simplifications that make this methodology computationally feasible, as we now discuss.

2.2. Application of Closure to the Min P Statistic

Given the typical genome scan, with each test yielding a p-value for genetic association, the first impulse is to locate the minimum value (min P). The question then becomes, "How unusual is the min P, given the number of genetic features scanned?" This question can be answered using an hypothesis testing approach, where one tests the global null hypothesis of no feature effect by evaluating the probability that the min P can be as low as the observed value, under the global null. This is similar to the approaches described in *(18,19)*, except that they do not apply the closure principle to isolate particular loci. Their analysis at the first step is essentially equivalent, but by applying the closure principle to their test procedure, one can obtain multiple candidate (single-level) associations between quantitative trait loci (QTLs) and the trait, all with familywise error protection, even under the case where there are some null and some non-null QTL locations.

Fortunately, one need not consider all 2^k-1 subsets for the closed procedure. If each subset is tested using min P from that subset, then one need only evaluate the k subsets that correspond to the ordered p-values, and not the entire set of 2^k-1. Formally, let the observed p-values be $p_1,\cdots,p_k$ ordered as $p_{(1)}\leq\cdots\leq p_{(k)}$ with corresponding hypotheses $H_{(1)},\cdots,H_{(k)}$ with $p_{(j)}=p_{i_j}$. Let the random p-values prior to observation be denoted by P_j.

The closed min P-based method collapses *(20)* to the following sequential procedure:

Algorithm: Closed min P Testing

reject $H_{(1)}$ if $P\,(\min_{j\in\{i_1,\dots,i_k\}}P_j\le p_{(1)})\le\alpha$

reject $H_{(2)}$ if $H_{(1)}$ was rejected and $P\,(\min_{j\in\{i_2,\dots,i_k\}}P_j\le p_{(2)})\le\alpha$

.

.

.

reject $H_{(k)}$ if $H_{(k-1)}$ was rejected and $P\,(\min_{j\in\{i_k\}}P_j\le p_{(k)})\le\alpha$

In accordance with the closure principle, all probabilities are calculated under the assumption of no genetic effect in the respective subsets of hypotheses.

Using the Bonferroni inequality

$$P\,(\min_{l\in\{i_j,\dots,i_k\}}P_l\le p_{(j)})\le(k-j+1)p_{(j)},$$

the closed min P-based procedure becomes the Holm method *(21)*. However, this method is needlessly conservative: the upper bound $(k-j+1)p_{(j)}$ is too large, implying that 5%-level significance might not be attained. This conservativeness arises because (1) there are correlations, sometimes large, among the genes due to linkage, and (2) the distributions of the tests are discrete *(22)*.

The correlation structure and discreteness of distributions can be taken into account by calculating the probabilities

$$P\,(\min_{l\in\{i_j,\dots,i_k\}}P_l\le p_{(j)})$$

directly and exactly using permutation tests. To do this, one randomly permutes the vectors of genetic indicators over the set of all subjects, so that in a given resampled data set, the first n_1 vectors are assumed to have phenotype 1, and the remaining n_2 are assumed to have phenotype 2. Thus, in this permutation model, the null hypothesis of no genetic effect holds for all subsets of hypotheses, as required by both the closure principle and the "subset pivotality" criterion of Westfall and Young *(23)*, p. 42. The probabilities

$$P\,(\min_{l\in\{i_j,\dots,i_k\}}P_l\le p_{(j)})$$

are then exactly computed as the proportion of possible permutations for which the value of $\min_{l\in\{i_j,\dots,i_k\}}P_l{}^*$, as calculated from the permuted data set, is less than or equal to the value $p_{(j)}$, as calculated from the original data set. Because resampled (or permuted) data sets preserve the correlation structure and discreteness characteristics, the resulting probabilities are typically less than the conservative Bonferroni approximations $(k-j+1)p_{(j)}$.

As the number of possible permutations can be exceedingly large, a simple and accurate approximation can be obtained by permutation resampling, that is, by sampling with replacement from the finite population of possible permutations. The resulting method is a statistically permutationally exact method under the case of infinitely many Monte Carlo samples. Monte Carlo error bounds and detailed algorithms are described by Westfall and Young *(23)*. Fortunately, software to perform this exact, closed min *P*-based analysis is readily available in PROC MULTTEST of SAS/STAT® *(24)*. This software requires a binary or ordinal phenotype such as (diseased)/(not diseased), or (severely diseased)/(moderately diseased)/(not diseased). The software runs more quickly when the phenotype is coded as binary. To take full advantage of the discreteness, it also requires binary genotype representations, although it can analyze ordinal genotype representations in exact fashion as well.

2.3. Application of Closure to the Simes–Hommel Test for Genetic Association

As an alternative to the use of the min *P* statistic for testing each subset homogeneity hypothesis, one may use Simes test *(25)*, which considers the entire distribution of *p*-values, rather than just the minimum. For a given set of *k* genetic association tests with *p*-values $p_1,...,p_k$, the hypothesis of no genetic effect is rejected if $\min\{kp_{(j)}/j\} \le \alpha$, where the $p_{(j)}$ are the ordered *p*-values. Like the case with closed testing and the min *P* test, closed testing with Simes' test allows shortcuts so that all $2^k - 1$ subsets need not be evaluated. The simplification occurs because, for each subset size (say, *s*), one need only consider the combined test that contains the gene of interest, and the $s - 1$ remaining largest *p*-values, rather than all $\binom{k}{s}$ subsets of size *s*. Hommel *(26)*, Wright *(27)*, and Grechanovsky and Hochberg *(28)* describe such shortcut methods. PROC MULTTEST of SAS/STAT® (as of Version 8.1) can perform these tests with $O(k^2)$ operations, rather than $O(2^k)$, which makes the method feasible for genetics screening tests.

The Simes test is valid (has type I error rate $\le \alpha$) when the tests are positively dependent, as shown by Sarkar *(29)*. In negatively dependent cases, the error rate may exceed α, but the excess is typically slight and not troubling *(30)*.

While it would be preferred to use the discreteness of the distributions for the Simes test *(31)*, as shown in **Subheading 2.2.** for the min *P* test, such an analysis would greatly increase the computational complexity. Studies have shown that the Simes-based approach tends to be more powerful than the min *P*-based approach when there are greater numbers of affected hypotheses *(32,33)*. In genetics experiments where multiple gene effects are expected, or with tight linkage, this might indeed be the case. In such a case, the Simes-based approach might have superior power to the min *P*-based approach. Further research is needed to develop computationally convenient Simes-based closed testing algorithms that incorporate distributional characteristics.

2.4. Application of Closure to the Fisher Test for Genetic Association

Yet another possibility is to apply the Fisher combination test *(34)* for each subset homogeneity hypothesis. For a given subset, the combination test statistic is $T = -2\Sigma \ln p_i$, which is distributed as χ^2_{2k} when (1) the subset homogeneity hypothesis is true, (2) the *p*-values are uniformly distributed, and (3) the tests are independent. Assumptions (2) and (3) are rather crucial here, but may be reasonable for the analysis of candidate genes that are expected to be only weakly linked, and when sample sizes are large. In gene expression tests, there is no linkage and the independence assumption might be more reasonable than in the case of gene–disease association tests.

Like the Simes-based tests, the Fisher combination-based test often allows several small *p*-values to reinforce one another to produce a more powerful test (than the min *P*-based method). The same $O(k^2)$ computational simplification seen for the closed Simes-based method described above holds for the closed Fisher combination method, making it also feasible for genetic association tests, and the method is available in PROC MULTTEST of SAS/STAT® (Version 8.1).

Pesarin *(35)* avoids the independence and uniformity assumptions, developing algorithms for exact Fisher combination tests that incorporate relevant distributional characteristics, including correlations. Further research is needed to develop computationally convenient closed testing algorithms that incorporate such tests.

3. Applications to Gene–Disease Associations

While one typically views the phenotype as a response (or penetrance) resulting from genetic predisposition, it is often reasonable (e.g., in case-control studies) to turn the problem on its head, and view genotype frequency as a function of the phenotype. In this section we apply the general closed testing methods described in **Subheading 2.** to specific genetic association tests, with the point of view of multiple comparisons of gene frequencies between cases and controls.

3.1. Multiple "Serological" Tests with Binary Phenotype

Consider the following 2×2 contingency table, cross-classifying disease status with presence of a particular allele at a given locus. The sample size is deliberately small to illustrate the main ideas.

Group	Allele A present	Allele A absent	Total
Case	5(100%)	0(0%)	5(100%)
Control	2(40%)	3(60%)	5(100%)

Sasieni *(8)* calls this a "serological" test because it "was common when HLA typing was done by serology, so that it was not possible to distinguish between [homozygous and heterozygous states]." He also notes that the resulting contingency table test (chi-square, χ^2) is completely efficient when allele A is dominant. Our analyses will consider the Fisher exact test instead of the χ^2 *(36)*.

Now consider the following arrangement of the contingency table in a "flat file" representation amenable to computer input.

Subject	Group	*D1*
01	Case	1
02	Case	1
03	Case	1
04	Case	1
05	Case	1
06	Control	1
07	Control	0
08	Control	1
09	Control	0
10	Control	0

Here, *D1* stands for "dominance coding at locus 1," and the 0s and 1s denote presence or absence of allele *A* at that locus. Now, in genomic scans, for example, using SNPs *(37)*, we will have multiple such indicators for a large collection of loci, resulting in a data set like that in **Table 1**, shown with just three loci for convenience.

For this data set, the Fisher exact (two-sided) *p*-values for testing associations between case-control status and locus are 0.1667, 1.0000, and 0.5238, respectively, for loci 1, 2, and 3. Nothing is significant, as expected with the small sample sizes; these values are used for illustration purposes only.

The closure principle described in **Subheading 2.1.** requires that additional *p*-values be computed for intersection hypotheses H_{12}: *D1* and *D2* are unaffected; H_{13}: *D1* and *D3* are unaffected; H_{23}: *D2* and *D3* are unaffected; and H_{123}: *D1*, *D2*, and *D3* are unaffected. By "unaffected" we mean that the distributions of the binary vectors are identical between Cases and Controls.

To calculate the exact closed min *P*-based multiple test procedure described in **Subheading 2.2.**, there are simplifications, and we require *p*-values only for the intersection hypotheses corresponding to the ordered *p*-values, H_{123}, H_{23}, and H_2. The *p*-value for H_{123} using the min *P* statistic is then $p_{123} = P(\min(P_1,P_2,P_3) \le 0.1667 | H_{123})$. To calculate this quantity exactly, one can enumerate all 10! permutations of the three-dimensional vectors, calculate $\min(P_1,P_2,P_3)$ for each permutation and note whether it is smaller than 0.1667, and take p_{123} to be the proportion of the 10! permutations (actually, only

Table 1
Input Form for Multiple Dominance Tests

Subject	Group	D1	D2	D3
01	Case	1	0	1
02	Case	1	0	1
03	Case	1	1	1
04	Case	1	0	0
05	Case	1	1	1
06	Control	1	0	1
07	Control	0	0	0
08	Control	1	1	1
09	Control	0	1	0
10	Control	0	0	0

$10!/[5!5!]$ are required) yielding a min P smaller than 0.1667. Alternately, one can sample from the permutation distribution. The following table shows one random sample from the multivariate permutation distribution:

Subject	Group	D1	D2	D3
07	Case	0	0	0
02	Case	1	0	1
10	Case	0	0	0
01	Case	1	0	0
08	Case	1	1	1
03	Control	1	1	1
09	Control	0	1	0
05	Control	1	1	1
04	Control	1	0	0
06	Control	1	0	1

For this sample, the p-values are, respectively, 1.0000, 0.5238, and 1.0000, with min P=0.5238. Thus, this is one of the 10! permutations for which min P is not smaller than 0.1667. Sampling all permutations, 21.43% of the permutations yield min P smaller than 0.1667, so $p_{123}=0.2143$ is the exact p-value for the composite H_{123} when the min P test is used. According to the closed min P testing algorithm in **Subheading 2.2.**, the hypothesis H_1 (which happens to correspond to the smallest p-value) would not be rejected, and no further inference could be made. However, if H_1 were rejected, then we could proceed to test H_3 using the p-value $p_{23}=P(\min(P_2,P_3)\leq0.5238|H_{23})$; H_3 would have been rejected if this probability were less than 0.05 (or whatever FWE level is chosen).

This analysis is automated in PROC MULTTEST of SAS/STAT®. The invoking code and testing portion of the output are as follows:

```
proc multtest data=table1 stepperm n=1000000 seed=121211;
  class group;
  test fisher(D1 D2 D3);
  contrast "compare" −1 1; run;
```

p-Values

Variable	Contrast	Raw	Stepdown Permutation
D1	Compare	0.1667	0.2149
D2	Compare	1.0000	1.0000
D3	Compare	0.5238	0.7855

The results of the closed testing algorithm are conveniently reported as adjusted *p*-values in the "Stepdown Permutation" column: if the adjusted *p*-value is <0.05, then the corresponding genetic association is significant at the FWE= 0.05 level using the closed min *P*-based testing algorithm of **Subheading 2.2.** Note also that the reported *p*-value 0.2149 differs slightly from the *p*-value 0.2143 obtained via direct enumeration of all 10! permutations; this difference reflects Monte Carlo error. As MULTTEST sampled 1,000,000 times, with replacement, from the population of permutations, the Monte Carlo standard error is just $\{0.2149(1-0.2149)/1000000\}^{1/2} = 0.00041$; thus the Monte Carlo estimate is 1.46 standard errors from the exact value, or acceptably close.

We have chosen 1,000,000 samples from the permutation distribution in this case, and the analysis takes less than a minute on a typical (as of the present date) PC workstation. In larger problems with more loci, it will take longer. We suggest at least 1000 samples to estimate the *p*-values with reasonable precision, although as large a number of samples as is convenient should ordinarily be chosen.

3.2. Testing Both Dominant and Recessive Modes of Inheritance

We may allow for recessive effects by considering 2×2 tables where genetic effect is coded as either (1) the gene is homozygous for the allele in question or (2) the gene is not homozygous for the allele in question. There is a high degree of dependence among such tests; this will be accommodated exactly in the closed multiple testing procedure. Following from **Table 1, Table 2** represents the input form suggested for such an analysis. Each gene has been coded two ways, with dominance coding *D* as shown in **Table 1**, and recessive coding *R*.

Note that there is positive correlation between the two codings, as a person who is "recessive" with respect to one allele is also "dominant" with respect to the other. There can also be strong positive correlation between closely linked genes owing to the linkage disequilibrium; nevertheless, these correlations are properly modeled via vector resampling as described previously.

Table 2
Dominant and Recessive Codings

Subject	Group	G1	D1	R1	G2	D2	R2	G3	D3	R3
01	Case	AA	1	1	aa	0	0	AA	1	1
02	Case	AA	1	1	aa	0	0	AA	1	1
03	Case	AA	1	1	AA	1	1	AA	1	1
04	Case	AA	1	1	aa	0	0	aa	0	0
05	Case	AA	1	1	Aa	1	0	AA	1	1
06	Control	Aa	1	0	aa	0	0	AA	1	1
07	Control	aa	0	0	aa	0	0	aa	0	0
08	Control	Aa	1	0	AA	1	1	Aa	1	0
09	Control	aa	0	0	AA	1	1	aa	0	0
10	Control	aa	0	0	aa	0	0	aa	0	0

Using the binary coding shown in **Table 2**, the specific hypotheses tested are $H_{0ij} : \pi_{1ij} = \pi_{2ij}$, where π_{1ij} denotes prevalence of coding i ($i=R,D$) for gene j ($j=1,2,3$) among controls; and where π_{2ij} denotes the corresponding quantity among cases. These hypotheses again are testable using the two-sided Fisher exact test, and the exact closed testing method is applicable as well. Code and output follow.

```
proc multtest data=table2 stepperm n=1000000 seed=121211;
   class group;
   test fisher(D1 R1 D2 R2 D3 R3);
   contrast "compare" –1 1; run;
```

p-Values

Variable	Contrast	Raw	Stepdown Permutation
D1	Compare	0.1667	0.3258
R1	Compare	0.0079	0.0161
D2	Compare	1.0000	1.0000
R2	Compare	1.0000	1.0000
D3	Compare	0.5238	0.7855
R3	Compare	0.2063	0.3490

There are several points to make about the results. First, the recessive genotype at locus 1 is considered statistically significant at the FWE = 0.05 level using the exact min P-based closed testing procedure, as the Stepdown Permutation p-value is < 0.05. Second, the adjustment of the unadjusted p-value 0.0079 to the adjusted 0.0161 is substantially less than one might expect with Bonferroni correction ($6 \times 0.0079 = 0.0474$); this savings comes as a result of using exact closed testing methods that incorporate discreteness as well as correlations. Third, it is somewhat unusual to find a more significant result when the family size is expanded, as we see here comparing the "dominance" analysis with three tests to the "dominance + recessive" analysis with six tests. However, when the expanded family contains tests that are more powerful, then it is certainly possible that there will be more significance in the expanded family, despite the larger multiple testing penalty. This example is suggestive of a situation where locus 1 has a purely recessive and fully penetrant effect.

This method can be extended to multiallelic genes as well. With multiple alleles the number of tests expands considerably: for $L > 2$ alleles, there will be $2L$ tests. (However, when $L = 2$ there are only two tests as, e.g., the dominant and recessive tests for allele a are completely determined by the corresponding tests for allele A.) Caution is recommended here, as large numbers of multiallelic genes can increase the family size substantially, thereby reducing power (in most cases).

3.3. Multiple Tests for Epistatic Effects

When two or more genes are necessary for the expression of the phenotype, we have an *epistasis*. It is thought that many complex traits and diseases are the result of the interaction of several rather common genotypes.

One possible method for screening gene combinations is to compare frequencies of the combinations occurring in either the case or the control populations. Let us revert to the "dominance" coding shown in **Table 1**, and consider whether combined effects of genes might signal differences in cases vs controls. The resulting data look like this:

Subject	Group	*D1*	*D2*	*D3*	*D1D2*	*D1D3*	*D2D3*
01	Case	1	0	1	0	1	0
02	Case	1	0	1	0	1	0
03	Case	1	1	1	1	1	1
04	Case	1	0	0	0	0	0
05	Case	1	1	1	1	1	1
06	Control	1	0	1	0	1	0
07	Control	0	0	0	0	0	0
08	Control	1	1	1	1	1	1
09	Control	0	1	0	0	0	0
10	Control	0	0	0	0	0	0

There are obviously correlations between the columns; in fact, in these data the *D1D3* column is identical to the *D3* column. The exact closed testing procedure automatically accounts for such dependencies, in effect reducing the multiplicative adjustment by one for each perfect dependency. The SAS code and output are as follows:

proc multtest data=table3 stepperm n=1000000 seed=121211;

class group;
test fisher(*D1 D2 D3 d1d2 d1d3 d2d3*);
contrast "compare" −1 1; run;

p-Values

Variable	Contrast	Raw	Stepdown Permutation
D1	Compar	0.1667	0.2864
D2	Compar	1.0000	1.0000
D3	Compar	0.5238	0.7855
D1D2	Compar	0.1667	0.2864
D1D3	Compar	0.5238	0.7855
D2D3	Compar	1.0000	1.0000

In this analysis, nothing would be considered significant, as none of the Stepdown Permutation (or closed exact Min *P* adjusted) *p*-values are less than the FWE 0.05 level. However, had there been a synergistic effect of two of these genes in dominant form, we might have seen some significant results.

One should be very cautious about using the multiplicative factors as shown here to discover epistatic effects; indiscriminant selection can greatly increase family size and thereby reduce power. For example, if there are 1000 genes, one might consider $1000(999)/2 = 499,500$ possible combinations. It is preferred to keep the family size smaller; thus this method is suggested when the number of genes is small, say 100 or less (assuming reasonable sample sizes in the case and control groups).

3.4. Stratification

One can conceive of several situations in which gene–disease associations should be analyzed using stratification. Two cases of major importance are as follows:

1. Epistasis involving a known gene. A known gene, say *G1*, contributes to disease. However, there are questions of epistasis concerning other genes. In this case, the epistatic effects should not be modeled as illustrated in **Subheading 3.3.** as the prevalence of *G1Gi* will surely differ between cases and controls, owing simply to the main effect of *G1*. In such a case it will be appropriate to compare the prevalence of genotype *Gi* among patients who share a common value of *G1*.

2. Environmental factors. An environmental factor, smoking, for example, might be a known contributor to disease. In such a case, it would be better to assess genetic contributions by partialling out the smoking variable, both to improve sensitivity of the tests and to remove a possible source of confounding.

Stratified analyses can be handled in an exact fashion, using essentially the same methods as described in **Subheadings 3.1.–3.3.,** but using exact stratified (Mantel–Haenszel) tests instead of Fisher exact tests. Exact *p*-values for these tests (analogous to the Fisher exact *p*-values) are easily obtained using existing software. The hypotheses differ depending on whether one is using stratified or unstratified analysis; with stratified analysis, the composite null hypothesis states that the distributions of the binary vectors are identical for both groups *within each stratum* (although the distributions are allowed to differ between strata). To test these hypotheses, we permute the observation vectors as before, but independently within strata.

The following invocation of PROC MULTTEST uses the data in **Table 1**, and treats *D3* as if it were a known gene contributing to disease (in dominant form), and performs exact, closed, stratified multiple testing.

```
proc multtest data=table1 stepperm n=1000000 seed=121211;
   class group;
   strata d3;
   test ca(D1 D2/permutation=20);
   contrast "compare" −1 1; run;
```

p-Values

Variable	Contrast	Raw	Stepdown Permutation
D1	Compare	0.2500	0.3993
D2	Compare	1.0000	1.0000

In this example we find no significant difference of *D1* frequency between cases and controls when analyzed within groups defined by *D3* status. Of course, this is a very small data set; in practice, we might apply this to hundreds of candidate genes, after stratifying on one known gene.

The syntax "ca" in the preceding SAS code stands for "Cochran–Armitage Trend test" *(38,39)*, which is equivalent to the Fisher exact test in the unstratified case, and which gives an exact stratified Fisher exact test in the stratified case. The syntax "permutation = 20" specifies exact permutation tests when the total number of observed G_i genotypes is < 20, or in this case, always. With large numbers of cases and controls, it is reasonable to specify "permutation = 100" or so to calculate exact permutation tests when the totals are < 100, but otherwise to use the normal approximation.

With sufficient sample size, this could be used as a forward stepwise procedure: select the most significant gene at step 1 (if significant by adjusted *p*-value); select the second major contributing gene (if significant by adjusted *p*-value), while partialling out the first as a "stratum" variable; select the third major contributing gene (if significant by adjusted *p*-value) while partialling the first and second selected variables as a combined "stratum" variable. This procedure has the attractive property that FWE is controlled at each stage, under the assumption of fixed ordering of variables. However, because the variables selected at earlier stages are random, it is possible that FWE is uncontrolled; *see* **ref.** *40* for further details in a related application.

Table 3
Ordinal Phenotype

Group	*D1*	*D2*	*D3*
Severe	1	0	1
Severe	1	0	1
Severe	1	1	1
Mild	1	0	0
Mild	1	1	1
None	1	0	1
None	0	0	0
None	1	1	1
None	0	1	0
None	0	0	0

3.5. Ordinal Phenotypes

Some phenotypic traits, for example, mental diseases, are best expressed ordinally, i.e., not diseased, mildly diseased, diseased, and badly diseased. One can perform a logistic regression of the binary genotype on the phenotypical outcome and test for significance of phenotype. The resulting logistic regression score test is equivalent to the Cochran–Armitage linear trend test that compares proportions of genotypes among the ordinal categories *(8)*.

In our paradigm of conditioning on the phenotype and examining the distribution of the genotypes, such an analysis can easily be accommodated as shown in **Subheadings 3.1.–3.4.**, with the exception that the tests are based on the exact permutation distribution of the (possibly stratified) Cochran–Armitage test instead of the Fisher exact two-sample tests. The null hypothesis for any given set of genotypes is that the multivariate binary genotype distribution is identical across all phenotype categories; or equivalently, that all permutations of the n multivariate binary vectors are equally likely. The trend test statistics are most sensitive to linear (or at least monotonic) departures from the null.

The exact closed min P-based analysis shown in the previous subheadings can be performed just as easily in this case. Consider the data in **Table 1**, but with the "Case/Control" variable recoded as "Severe," "Mild," and "None" **(Table 3)**. The following SAS code and output show how to perform the exact, closed multiple testing procedure with these data.

```
proc multtest data=table4 order=data stepperm n=1000000 seed=121211;
   class group;
   test ca(D1 D2 D3/permutation=20);
   contrast "trend" –1 0 1; run;
```

p-Values

Variable	Contrast	Raw	Stepdown Permutation
D1	Trend	0.1417	0.2349
D2	Trend	1.0000	1.0000
D3	Trend	0.1667	0.3108

Here the "raw" p-values are exact Cochran–Armitage permutation p-values, and the Stepdown Permutation p-values are obtained by evaluating the distribution of min P over subsets corresponding to the ordered raw p-values, as described in **Subheading 3.1.**

Trend tests can be more powerful than Fisher exact tests that collapse the phenotype into two categories (e.g., here we might classify both "Severe" and "Mild" into "Case," and leave "None" as "Control.") However, one should limit the number of categories (say, to five or fewer); otherwise, the tables can become sparse and the tests can lose power. (Note: the MULTTEST procedure also computes exact stratified Cochran–Armitage trend tests for ordinal phenotype, as shown in **Subheading 3.3.** for the case of binary phenotype.)

3.6. Ordinal Genotypes: Cumulative Polygenic Effects

Suppose the disease is associated with allele A on a biallelic gene. In some disease models, the effect or penetrance of the gene is higher for heterozygotes than for homozygotes aa, while the effect or penetrance of the gene is still higher for homozygotes AA than for heterozygotes. This suggests a linear model relating phenotype Y to the ordinal genotype X, where $X=0$, 1, or 2 for genotypes aa, Aa, and AA, respectively. Assuming Y is binary or ordinal as we have done, and again turning the problem on its head, we can compare the distribution of X for the different categories of Y, and test for trend, much as with the Cochran–Armitage test. Exact nonparametric tests for trend are available in, for example, StatXact *(41)*, so, in principle, the problem of exact closed multiple tests is solved for this case as well. However, there is no ready-made software for this purpose. A reasonable solution is available in PROC MULTTEST where one uses the parametric test for genotype i to obtain p-values p_i, then finds the multiplicity-adjusted p-values $P(\min P_j \le p_i)$. The final result is obtained by permuting vectors as before, so the final analysis is exact (modulo Monte Carlo error, which can be reduced to an arbitrarily low level). Using the parametric unadjusted p-values in the exact min P test can cause problems of imbalance *(42,43)*; however, this problem is often not a major issue *(44)*.

Another source of ordinal variables that can be handled similarly is the cumulative effect of several genes. For example, it may be thought that the

Table 4
Ordinal Genotypes

Subject	Group	G1	X1	G2	X2	G3	X3	X4
01	Case	AA	2	aa	0	AA	2	4
02	Case	AA	2	aa	0	AA	2	4
03	Case	AA	2	AA	2	AA	2	6
04	Case	AA	2	aa	0	aa	0	2
05	Case	AA	2	Aa	1	AA	2	5
06	Control	Aa	1	aa	0	AA	2	3
07	Control	aa	0	aa	0	aa	0	0
08	Control	Aa	1	AA	2	Aa	1	4
09	Control	aa	0	AA	2	aa	0	2
10	Control	aa	0	aa	0	aa	0	0

alleles A_1, A_2, and A_3 (in genes 1, 2, and 3, respectively) contribute cumulatively to the phenotype Y. In this case the variable $X_4 = X_1 + X_2 + X_3$ is suggested. Other codings are possible, such as $X_4 = I(X_1 > 0) + I(X_2 > 0) + I(X_3 > 0)$, where $I(\bullet)$ denotes the indicator function, as would be suggested if the disease is cumulatively related to dominant expressions of the A_i only, with no extra "bump" for recessivity. Note also that such a combination presupposes that the directions of allelic associations are known for all genotypes, which might be rare. Nevertheless, the method is shown below to illustrate the possibility, and to note that perfect dependencies of the type induced by the X_4 variable cause no problems with the exact min P-based testing method.

The data in **Table 4** relate directly to **Table 2**, with $X_1 - X_4$ as just described. Exact closed multiple testing is accomplished via the following code, and the results are shown as follows:

```
proc multtest data=table5 stepperm n=1000000 seed=121211;
  class group;
  test mean(X1–X4);
  contrast "compare" –1 1; run;
```

p-Values

Variable	Contrast	Raw	Stepdown Permutation
X1	Compare	0.0002	0.0161
X2	Compare	0.7599	1.0000
X3	Compare	0.1151	0.3490
X4	Compare	0.0497	0.1272

The result here is that the gene G1 ordinal variable has a different mean for the two phenotypes, as its adjusted p-value is < 0.05.

In this analysis the "Raw" p-values are inexact, being based on the normal-theory t-test, but the Stepdown Permutation p-values are exact modulo Monte Carlo error. The inexactness of the Raw p is suggested by the fact that it is so small, relative to the exact multiplicity-adjusted p-value.

4. Application to a Large Simulated Data Set

The data for this study (ftp://statgen.ncsu.edu/pub/zaykin/cand/) were simulated according to the following model. We simulated a genetic map of 20 candidate regions. Each 100-kb candidate region contained 10 uniformly, randomly spaced SNPs. Candidate regions themselves were assumed unlinked; however, the recombination process for SNPs inside candidate regions was modeled directly, assuming Haldane's mapping function (no interference) and Poisson-distributed number of recombination events with mean equal to the genetic length in Morgans. Three of 10 candidate regions contained disease genes. Four SNPs in each of first two regions and three SNPs in the third region were assumed to be contributing to the disease.

We used an additive model with weak interaction to model penetrances. According to this model, one allele for each of 11 SNPs was assigned a uniform random genetic effect, additively contributing to the total probability of developing disease (genetic penetrance), but the final penetrance for each genotype class was given a 0–5% uniform random deviation. Finally, $3^{11} = 177,147$ individual penetrances for individual multilocus genotypes were scaled between 0 and 1, so that the "typical" penetrance value of a multilocus genotype was about 50%.

We allowed for separate sexes, with no selfing, and no allowance of sib matings. Generations were assumed to be discrete. We simulated five originally homogeneous equilibrium populations of 500 individuals each, and allowed for 100–200 generations of genetic drift with population growth rate of 1.2, and a migration rate of 0.2 from each of the four populations into the fifth during the first 35 generations. The maximum population size was set to 15,000 individuals. We kept only populations with the final disease prevalence in the range 5–15%. We sampled 500 of affected and 500 of nonaffected individuals from the admixed population at the final generation.

To illustrate the method we simulated sets of data using two different models. The first model (model 1) is as described in the preceding. The second model (model 2) differs in that the chromosome regions are themselves closely linked, so that the association may extend over all 20 regions.

Table 5 contains part of the resulting analysis of a typical data set (20 smallest p-values) simulated under the first model, and **Table 6** contains part of the analysis for data that were simulated under the second model (200 generations). The

Table 5
p-Values for Simulated Data, Model 1

Genotype	Unadjusted p-value	Closed Bonferroni	Closed Sidak	Closed Permutation
D7	0.0000000	0.00000	0.00000	0.000
D9	0.0000000	0.00000	0.00000	0.000
R9	0.0000000	0.00000	0.00000	0.000
D10	0.0000000	0.00000	0.00000	0.000
R10	0.0000000	0.00000	0.00000	0.000
R7	0.0000006	0.00019	0.00019	0.000
R5	0.0000198	0.00695	0.00692	0.005
D5	0.0000723	0.02548	0.02515	0.015
D8	0.0002896	0.10276	0.09767	0.056
R132	0.0004779	0.16953	0.15597	0.091
R17	0.0008843	0.31228	0.26832	0.201
D133	0.0019127	0.67437	0.49084	0.384
R3	0.0023851	0.83925	0.56838	0.452
D127	0.0024807	0.86887	0.58101	0.462
D70	0.0026634	0.92573	0.60423	0.482
D113	0.0034234	1.00000	0.69534	0.572
D136	0.0034387	1.00000	0.69570	0.572
R8	0.0039253	1.00000	0.74653	0.635
R172	0.0046352	1.00000	0.80447	0.693
D129	0.0095221	1.00000	0.96444	0.915

analysis for both tables was performed using PROC MULTTEST, Version 8.1 (an example of the invoking program code is given in the Appendix).

Actual regions contributing to the probability of developing the disease were typed with markers labeled 1–15, so the algorithm correctly identifies SNPs typed in all three regions. Originally small p-values corresponding to the false regions become nonsignificant after proper multiplicity adjustment over the set of 400 tests. Note that closed permutation p-values are smaller than the closed Bonferroni and Sidak (independence-assuming) corrections. The effect is more pronounced when long regions of densely mapped SNPs are considered **(Table 6)**. For example, the closed Bonferroni-adjusted p-value for R134 is 0.0089834, but corresponding exact (modulo Monte Carlo error) min P permutation adjustment is 0.002.

5. Application to Gene Expression Data

Gene expression data may be analyzed using similar techniques. Data given in Golub et al. *(45)* are available at http://waldo.wi.mit.edu/MPR/ data_set_ALL_AML.html) for relating gene expression from 7129 genes to

Table 6
P-Values for Simulated Data, Model 2

Genotype	Unadjusted p-value	Closed Bonferroni	Closed Sidak	Closed Permutation
R138	0.000014178	0.0043713	0.0043618	0.001
D42	0.000014739	0.0045309	0.0045207	0.001
R83	0.000017126	0.0052764	0.0052625	0.002
R115	0.000018588	0.0057627	0.0057462	0.002
R178	0.000022408	0.0068857	0.0068621	0.002
D194	0.000026979	0.0082854	0.0082512	0.002
D100	0.000028380	0.0087390	0.0087011	0.002
R134	0.000029131	0.0089834	0.0089432	0.002
D6	0.00003150	0.009618	0.009572	0.003
R191	0.00003233	0.009875	0.009827	0.004
R149	0.00003882	0.011811	0.011741	0.005
R73	0.00003975	0.012020	0.011949	0.006
D123	0.00004579	0.013792	0.013697	0.007
R100	0.00006178	0.018604	0.018433	0.009
D130	0.00006465	0.019422	0.019236	0.009
D21	0.00006529	0.019568	0.019378	0.009
R22	0.00007886	0.023409	0.023138	0.010
R104	0.00008572	0.025422	0.025102	0.011
D177	0.00008980	0.026545	0.026196	0.013
R173	0.00010426	0.030858	0.030388	0.015

disease status. (Golub et al. *[45]* apparently consider only 6817 of the 7129 available on the data set.) There are 11 patients with acute myeloid leukemia (AML) and 27 with acute lymphoblastic leukemia (ALL).

As discussed in **ref. *45***, one goal is to discriminate between the known AML and ALL populations on the basis of the observable gene expressions. Discriminant analysis (DA) is commonly used for this purpose, and a first step in DA is often to test for differences between the groups using Hotelling's T^2 test *(46)*. However, the T^2 test requires a nonsingular covariance matrix, and in this case the 7129×7129 covariance is quite singular, having rank somewhere near $11 + 27 = 38$, and the test cannot be applied. Nevertheless, the min P test can be carried out exactly to test for global differences; in addition, the exact min P-based closed testing procedure allows one to specify particular genes where simple associations exist, with full FWE protection.

In the gene expression data the response variable is continuous, and exact, distribution-free closed testing methods are available, as described in **Subheading 3.5.** One may test a global hypothesis using the max T statistic, where T is calculated as

$$T = \frac{\bar{X}_{ALL} - \bar{X}_{AML}}{s_p\sqrt{1/11 + 1/27}}$$

where $\bar{X}_{ALL}$ and $\bar{X}_{AML}$ refer to average expression in the ALL and AMR groups, and s_p is the pooled standard deviation. However, because max T is monotonically related to min P, where the p-values are calculated using the t-distribution with df $= 11 + 27 - 2$, this method is exactly equivalent to the min P testing method described in **Subheading 3**. PROC MULTTEST accomplishes this by resampling the 7129-dimensional vectors of gene expressions without replacement into like data sets having 27 ALL and 11 AML patients, then recomputing max T^* for the resampled data set. The p-value for the appropriate intersection hypothesis is then reported as the proportion of resampled data sets yielding max T^* greater than the original observed max T.

Golub et al. *(45)* performed a related permutation based-analysis using the statistic $T' = (\bar{X}_{ALL} - \bar{X}_{AML})/(s_1 + s_2)$. It would be equally possible to perform the exact closed testing procedure max T' as the base test, if desired. The benefits of the MULTTEST analysis are that (1) it is easily available and (2) it is known to control FWE via the closure principle.

Note that, as in the case of gene–disease tests, vector correlations are incorporated via vector resampling. However, in the case of gene expression data, there is no linkage, and therefore large correlations are not expected. Nevertheless, there is sample-specific dependence because the number of variables far exceeds the number of observations. This dependence is used to reduce the p-values, legitimately, because the tests are exact. Furthermore, as noted previously, the sample 7129×7129 covariance matrix among the gene expressions is massively singular, but this poses no difficulties whatsoever; the exact multiple testing procedure legitimately incorporates such sample-specific dependencies into the multiplicity adjustments via vector permutation resampling.

Such a test is an exact permutation test when all $\binom{38}{27}$ distinct resampled data sets (more than a billion) are enumerated. However, a reasonable approximation can be obtained by sampling randomly and with replacement from that set of permutations, and this is the PROC MULTTEST approach for testing global hypotheses. To make inferences about the specific genes, the closure method is used, and once again, only the subsets corresponding to the 7129 ordered p-values need to be evaluated, not the entire set of 2^{7129} subsets. Thus, once again, the MULTTEST procedure provides a closed testing method that is computationally feasible.

Table 7
***p*-Values for Golub Leukemia Data Set**

Gene	Unadjusted *p*-value	Bonferroni–Holm	Closed Min *P*
GENE3320	1.3824E-10	0.000001	0.0001
GENE4847	2.4355E-10	0.000002	0.0001
GENE2020	6.578E-10	0.000005	0.0001
GENE1745	0.000000010	0.000070	0.0004
GENE5039	0.000000010	0.000072	0.0004
GENE1834	0.000000015	0.000108	0.0005
GENE461	0.000000036	0.000257	0.0005
GENE4196	0.000000062	0.000438	0.0009
GENE3847	0.000000072	0.000510	0.0010
GENE2288	0.000000089	0.000635	0.0011
GENE1249	0.000000174	0.001239	0.0017
GENE6201	0.000000176	0.001250	0.0017
GENE2242	0.000000195	0.001386	0.0020
GENE3258	0.000000211	0.001500	0.0021
GENE1882	0.000000319	0.002267	0.0024
GENE2111	0.000000366	0.002606	0.0027
GENE2121	0.000000578	0.004115	0.0041
GENE6200	0.000000623	0.004428	0.0042
GENE6373	0.000000819	0.005823	0.0058
GENE6539	0.000001120	0.007961	0.0082
GENE2043	0.000001260	0.008954	0.0092
GENE2759	0.000001309	0.009304	0.0092
GENE6803	0.000001429	0.010156	0.0101
GENE1674	0.000001480	0.010519	0.0103
GENE2402	0.000001523	0.010821	0.0107
GENE2186	0.000001657	0.011770	0.0111
GENE6376	0.000002092	0.014856	0.0142
GENE3605	0.000002553	0.018133	0.0157
GENE6806	0.000002584	0.018352	0.0159
GENE1829	0.000002727	0.019364	0.0168
GENE6797	0.000003014	0.021399	0.0180
GENE6677	0.000003439	0.024412	0.0196
GENE4052	0.000003701	0.026268	0.0220
GENE1394	0.000004925	0.034948	0.0282
GENE6405	0.000005353	0.037980	0.0300
GENE248	0.000006381	0.045267	0.0346
GENE2267	0.000006488	0.046019	0.0352
GENE6041	0.000007802	0.055335	0.0421
GENE6005	0.000008019	0.056861	0.0428
GENE5772	0.000008994	0.063771	0.0471
GENE6378	0.000009591	0.067993	0.0500

The results are given in **Table 7** and the invoking MULTTEST code is given in the Appendix. Surprisingly, several results are significant, despite the small sample sizes and large degree of multiplicity. The association of leukemia subtype with the expression phenotype is confirmed; tests with closed permutation-based adjusted p-values < 0.05 indicate significant associations at the 0.05 FWE level.

Also note that 10,000 samples are generated from the permutation distribution, and all 7129 ordered tests were processed for each sample. This took only 20 min on a Windows NT workstation.

The effect of incorporating the sample-specific dependencies among the p-values is not as great as one might hope with such massive singularity in the covariance matrix. Naively, one might expect (or hope) that the effective Bonferroni multiplier would be on the order of $38 = 27 + 11$, the approximate rank of the 7129×7129 covariance matrix, when the dependence structure is incorporated correctly. However, this is not so. Dividing the adjusted p-values by the unadjusted p-values gives the effective multipliers; for example, the effective multiplier for the test involving "GENE248" is $0.0445267/0.000006381 = 7094.0$ for the Bonferroni–Holm procedure, but only $0.0346/0.000006381 = 5422.2$ for the exact min P-based closed procedure. The savings from using the correlation structure is to reduce the multipliers some, but not nearly to the extent suggested by the rank of the covariance matrix.

The Simes–Hommel method described in **Subheading 2.3.** also was applied to these data; the results were almost identical to Bonferroni–Holm, but the run took nearly 24 h because of the large number of tests.

Finally, we note that there are occasionally extreme outliers in the gene expression data. The negative effects of outliers can be diminished through log transformation as in **ref. 45** or one can use the rank transformation to avoid taking the logarithm of numbers that are less than or equal to zero. Use of the rank transformation in conjunction with permutation resampling in PROC MULTTEST provides an exact permutation-based closed testing procedure as before. However, this procedure is also attractive because the marginal tests are approximately valid rank-based permutation tests as well, being based on the rank transform *(47)*. When the analysis is performed on the rank-transformed expression data, there are a few changes in **Table 7**, mostly additions of variables where a large outlier masked the difference using the two-sample t-test.

Appendix

The following SAS/STAT® code was used to produce the results shown in **Tables 5** and **6**. It is assumed that the input file (in this case "SNP.DAT") has the result of each gene test coded as $AA = 1\ 1$, $Aa = 1\ 0$, $aA = 0\ 1$, and $aa = 0\ 0$;

and has the binary phenotype in the first column. The macro "%trans" recodes these data into "*AA* vs not *AA*" and "*aa* vs not *aa*" categories.

```
%macro trans;
   %do i = 1 %to 200;
   %let i1 = %eval(2*&i-1);
   %let i2 = %eval(2*&i);
   d&i = (bin&i1+bin&i2)=0;
   r&i = (bin&i1+bin&i2)=2;
   %end;
%mend;
data snp;
   infile "snp.dat" lrecl=10000;
   input y bin1-bin400;
   %trans;
   keep y d1-d200 r1-r200;
   run;
proc multtest data=snp stepbon stepsid noprint out=pval stepperm n=10000;
   class y;
   test ca(d1-d200 r1-r200/permutation=100);
   contrast "dis v nondis" 0 1;
run;
proc sort data=pval;
   by raw_p;
proc print data=pval;
   var _var_ raw_p stpbon_p stpsid_p stppermp;
   where raw_p<.05;
run;
```

The following SAS/STAT® code was used to analyze the data shown in **Table 7**. It is assumed that the SAS data set ("gene.express") has the result of each gene expression test in variable GENEi, and that the treatment indicators (AML or ALL) are contained in the variable called "disease."

```
proc multtest data = gene.express out=adjp stepperm holm n=10000 noprint;
   class disease;
   test mean(gene1-gene7129);
   contrast "AML vs ALL" -1 1;
run;
proc sort data=adjp(where=(stppermp le .05));
   by raw_p;
proc print data=adjp(where=(stppermp le .05)) noobs label;
   var _var_ raw_p stpbon_p stppermp;
run;
```

References

1. Bevan, S., Popat, S., and Houlston, R. S. (1999) Relative power of linkage and transmission disequilibrium test strategies to detect non-HLA linked coeliac disease susceptibility genes *Gut* **45,** 668–671.

2. Barnes, K. C. (1999) Gene-environment and gene-gene interaction studies in the molecular genetic analysis of asthma and atopy. *Clin. Exp. Allergy* **29** (Suppl 4)**,** 47–51.

3. El-Gabalawy, H. S., Goldbach-Mansky, R., Smith, D., Arayssi, T., Bale, S., Gulko, P., et al. (1999) Association of HLA alleles and clinical features in patients with synovitis of recent onset. *Arthrit. Rheum.* **42,** 1696–1705.

4. Tomer, Y., Barbesino, G., Greenberg, D. A., Concepcion, E., and Davies, T. F. (1999) Mapping the major susceptibility loci for familial Graves' and Hashimoto's diseases: evidence for genetic heterogeneity and gene interactions. *J. Clin. Endocrinol. Metab.* **84,** 4656–4664.

5. Wicker, L. S., Todd, J. A., and Peterson, L. B. (1995) Genetic control of autoimmune diabetes in the nod mouse. *Annu. Rev. Immunol.* **13,** 179–200.

6. Bodmer, W. F. (1986) Human genetics: the molecular challenge. *Cold Spring Harbor Symp. Quant. Biol.* **51,** 1–13.

7. Hagmann, M. (1999) A good SNP may be hard to find. *Science* **285,** 21–22.

8. Sasieni, P. D. (1997) From genotypes to genes: doubling the sample size. *Biometrics* **53,** 1253–1261.

9. Chiano M. N. and Clayton D. G. (1998) Fine genetic mapping using haplotype analysis and the missing data problem. *Ann. Hum. Genet.* **62,** 55–60.

10. Miller, R. C. (1981) *Simultaneous Statistical Inference*, 2nd edit. Springer-Verlag, New York.

11. Smouse, P. E. and Williams, R. C. (1982) Multivariate analysis of HLA–disease association. *Biometrics* **38,** 757–768.

12. Lander, E. and Kruglyak, L. (1995) Genetic dissection of complex traits: guidelines for interpreting and reporting linkage results. *Nat. Genet.* **11,** 241–247.

13. Zaykin, D. V., Zhivotovsky, L. A., Weir B. S., and Westfall, P. H. (2000) Truncated product method for combining *p*-values. Unpublished manuscript.

14. Weller, J. I., Song, J. Z., Heyen, D. W., Lewin, H. A., and Ron, M. (1998) A new approach to the problem of multiple comparisons in the genetic dissection of complex traits. *Genetics* **150,** 1699–1706.

15. Benjamini, Y. and Hochberg, Y. (1995) Controlling the false discovery rate — a practical and powerful approach to multiple testing. *JRSS-B* **57,** 289–300.

16. Zaykin, D. V., Young, S. S., and Westfall, P. H. (2000) Using the false discovery rate approach to the genetic dissection of complex traits: a response to Weller et al. *Genetics* **154,** 1917–1918.

17. Marcus, R., Peritz, E., and Gabriel, K. R. (1976) On closed testing procedures with special reference to ordered analysis of variance. *Biometrika* **63,** 655–660.

18. Churchill, G. A. and Doerge, R. W. (1994) Empirical threshold values for quantitative trait mapping. *Genetics* **138,** 963–971.

19. Doerge, R. W. and Churchill, G. A. (1996) Permutation tests for multiple loci affecting a quantitative character. *Genetics* **142,** 285–294.

20. Westfall, P. H. and Wolfinger, R. D. (2000) Closed Multiple Testing Procedures and PROC MULTTEST. SAS Observations, http://www.sas.com/service/library/periodicals/obs/observations.html.

21. Holm, S. (1979) A simple sequentially rejective multiple test procedure. *Scand. J. Statist.* **6**, 65–70.

22. Westfall, P. H. and Wolfinger, R. D. (1997) Multiple tests with discrete distributions. *Am. Stat.* **51**, 3–8.

23. Westfall, P. H. and Young, S.S. (1993) *Resampling-Based Multiple Testing: Examples and Methods for p-Value Adjustment.* John Wiley & Sons, New York.

24. SAS Institute Inc. (1999) SAS OnlineDoc ®, Version 8, Cary, NC: SAS Institute Inc.

25. Simes, R. J. (1986) An improved Bonferroni procedure for multiple tests of significance. *Biometrika* **73**, 751–754.

26. Hommel, G. (1988) A stagewise rejective multiple test procedure based on a modified Bonferroni test. *Biometrika* **75**, 383–386.

27. Wright, S. P. (1992) Adjusted *p*-values for simultaneous inference. *Biometrics* **48**, 1005–1014.

28. Grechanovsky, E. and Hochberg, Y. (1999) Closed procedures are better and often admit a shortcut. *J. Stat. Plan. Infer.* **76**, 79–91.

29. Sarkar, S. (1998) Some probability inequalities for ordered MTP_2 random variables: a proof of the Simes conjecture. *Ann. Statist.* **26**, 494–504.

30. Sarkar, S. and Chang, C. K. (1997) Simes' method for multiple hypothesis testing with positively dependent test statistics. *JASA* **92**, 1601–1608.

31. Krummenauer, F. and Hommel, G. (1999) The size of Simes' global test for discrete test statistics. *J. Stat. Plan. Infer.* **82**, 151–162.

32. Dunnett, C. W. and Tamhane, A. C. (1993) Power comparisons of some step-up multiple test procedures. *Statist. Prob. Lett.* **16**, 55–58.

33. Dunnett, C. W. and Tamhane, A. C. (1995) Step-up multiple testing of parameters with unequally correlated estimates. *Biometrics* **51**, 217–227.

34. Fisher, R. A. (1932) *Statistical Methods for Research Workers.* Oliver and Boyd, London.

35. Pesarin, F. (1999) *Permutation Testing of Multidimensional Hypotheses by Nonparametric Combination of Dependent Tests.* CLEUP University Publisher, Padova.

36. Weir, B. S. (1996) *Genetic Data Analysis II.* Sinauer Associates, Sunderland, MA.

37. Martin, E. R., Lai, E. H., Gilbert, J. R., Rogala, A. R., Afshari, A. J., Riley, J., et al. (2000.) SNPing away at complex diseases: analysis of single-nucleotide polymorphisms around APOE in Alzheimer disease. *Am. J. Hum. Genetics* **67**, 383–394.

38. Cochran, W. (1954) Some methods for strengthening the common χ^2 tests. *Biometrics* **10**, 417–451.

39. Armitage, P. (1955) Tests for linear trend in proportions and frequencies. *Biometrics* **11**, 375–386.

40. Westfall, P. H., Young, S. S., and Lin, D. K. J. (1997) Forward selection error control in the analysis of supersaturated designs. *Statist. Sinica* **8**, 101–117.

41. Mehta, C. and Patel, N. (1998) *StatXact: statistical software for exact non-parametric inference.* CYTEL Software, Cambridge, MA.

42. Beran, R. (1988) Balanced simultaneous confidence sets. *JASA* **83,** 679–686.

43. Beran, R. (1988) Prepivoting test statistics: a bootstrap view of asymptotic refinements. *JASA* **83,** 687–697.

44. Tu, W. and Zhou, X. H. (2000) Pairwise comparisons of the means of skewed data. *J. Stat. Plan. Infer.* **88,** 59–74.

45. Golub, T. R., Slonim, D. K., Tamayo, P., Huard, C., Gaasenbeek, M., Mesirov, J. P., et al. (1999) Molecular classification of cancer: class discovery and class prediction by gene expression monitoring. *Science* **286,** 531–537.

46. Johnson, R. A. and Wichern, D. W. (1998) *Applied Multivariate Statistical Analysis*, 4th edit. Prentice Hall, Englewood Cliffs, NJ.

47. Conover, W. J. and Iman, R. L. (1981) Rank transformation as a bridge between parametric and nonparametric statistics. *Am. Statist.* **35,** 124–129.

9

Statistical Considerations in Assessing Molecular Markers for Cancer Prognosis and Treatment Efficacy

James Dignam, John Bryant, and Soonmyung Paik

1. Introduction

The development and growth of molecular biologic technology is leading to a new appreciation of inherent heterogeneity in cancer. While long appreciated as morphologically diverse entities, malignancies have increasingly been explored for molecular characteristics indicative of cellular regulation and growth, ability to adapt to and change local environments, and susceptibility to potentially therapeutic agents. These pursuits have led to important advances in the understanding of cancer biology, and in selected instances, have led to the development and use of treatments designed to act on molecular targets.

Depending on the technology used to obtain the data, the evaluation of molecular disease characteristic markers in relation to outcomes may involve novel statistical analysis problems, as well as familiar design and analysis issues. The recent introduction of DNA microarray technology, in which dozens or even hundreds of molecular characteristics of a tumor can be quantified and compared to normal tissue or to other tumors, is a relevant example. Researchers are interested in which of these molecular markers may be indicative of poorer outcomes or response to specific therapies. An appropriate evaluation of this extremely large volume of data challenges the limits of current statistical methodology.

In this chapter, we examine a current research question in breast cancer biology as an illustrative example to circumscribe methods for the analysis of new molecular markers in relation to clinical outcome data. Specifically, the clinical utility of a molecular characteristic of breast cancer tumors is evaluated, using archived tumor samples combined with clinical follow-up information

From: *Methods in Molecular Biology, vol. 184: Biostatistical Methods*
Edited by: S. W. Looney © Humana Press Inc., Totowa, NJ

collected from a randomized clinical trial. This marker, the overexpression of the erbB-2 (also referred to as HER2/neu) protein on the cell surfaces of breast tumors, can potentially be used to select which chemotherapy drugs are liable to be of most benefit, and also has led to the development of new treatment agents designed to target the growth factor receptor encoded by the *erbB-2* oncogene.

2. Prognostic and Predictive Markers In Cancer

To better anticipate outcomes and tailor treatment for individuals with cancer, factors potentially indicative of prognosis have been investigated and employed in clinical decision-making. The extent of disease development and spread at time of diagnosis, usually a composite of features collectively referred to as the stage, is an important prognostic factor in all cancers. Related characteristics, such as size of the tumor, as well as the predominant tumor cell type and other pathologic features, are also well-recognized indicators of prognosis. Additional specific tumor cell characteristics, including the expression of receptors and protein complexes on tumor cell surfaces and the presence of genes in mutated form, may be associated with poor prognosis and/or poor response to treatment.

On this latter note, an important concept popularized in cancer studies but possibly unknown to statisticians (or known by another name) relates to factors that predict differential response to therapy in absence of or in addition to any relationship to prognosis in general. The term *prognostic factor* is generally reserved for those factors that identify patients at increased risk of relapse or death, as in the case of stage mentioned above. Factors that preferentially identify patients who respond to a given treatment are referred to as *predictive factors*. For example, tumors with certain pathologic characteristics may appear insensitive to chemotherapy. A characteristic can be both prognostic and predictive, an example being estrogen receptors found on the surface of breast tumor cells. The absence of such receptors is indicative of loss of cellular regulation and generally more profound pathologic aberrations leading to poorer outcomes. Furthermore, those patients with estrogen-receptor-bearing tumors have been found to be amenable to treatment with tamoxifen, an estrogen-like compound that blocks receptors and inhibits cell growth. Thus, estrogen receptors are both a prognostic marker and are predictive of treatment response with a specific, targeted agent. In this discussion, we will generally refer to markers under evaluation as prognostic markers, with the understanding that such markers will be evaluated for any relationship to treatment efficacy as well. Factors that might segregate patients who do not require chemotherapy after surgery from among those with early stage breast cancer are of particular interest, as there remains considerable debate regarding the worth of such treatment among so-called "good risk" patients *(1–3)*.

Clinical utility of a marker is generally defined as the circumstance whereby knowledge of the marker value can prompt clinical action, including increased diagnostic vigilance or specific treatment administration, which may benefit the patient. Because there has been a proliferation of potential markers in cancer, and yet little progress in achieving clinical utility for most, efforts have been introduced to define guidelines and criteria for new marker evaluation. The College of American Pathologists (CAP) has recently defined a three-level category ranking system for prognostic factors *(4)*. Expert panels comprised of pathologists, cancer biologists, clinicians, statisticians, and others have been convened periodically to deliberate on the substance and quality of evidence for prognostic factors in cancer. CAP category I factors have proven value established in several studies, preferably prospective trials where marker evaluation was a study objective. Category II factors are those with evidence of utility that require further study and verification. Category III factors are generally new markers with limited data available thus far. These include anecdotal and small data observations, usually accompanied by a substantive underlying biological motivation.

The concept of level of evidence (LOE) has been established for the evaluation of data concerning therapeutic interventions (described at the website http://cancernet.nci.nih.gov/clinpdq). It has been proposed that a similar scheme be applied to prognostic marker studies, so that physicians and other scientists can more uniformly and objectively evaluate the literature and better develop a research agenda to address outstanding questions. **Table 1** shows the LOE evaluation criteria suggested by Hayes and colleagues as part of a comprehensive system to evaluate markers for clinical utility *(5)*. Their TMUGS (Tumor Marker Utility Grading System) was developed in response to the somewhat haphazard manner in which marker information has developed over time, contributing to the relatively small improvement in prospective clinical evaluation of cancer patients. The LOE scale is applied to available clinical studies and, using this information together with an assessment of the assay methods, a semiquantitative score is derived reflecting to what extent evaluation of patients for the marker should become part of routine clinical decision-making.

It should be noted that, while a study satisfying the CAP category I or LOE I criteria would be ideal for unequivocally establishing the role of a new marker, such studies are unlikely to be conducted. The financial resources available for studies focused on markers rather than potentially therapeutic interventions are limited, and there are ethical implications of increasing sample size for therapeutic clinical trials to accommodate adequately powered ancillary studies of prognostic markers. Despite these barriers to the conduct of optimally designed marker studies, there is substantial opportunity for improvement of such studies within practical limitations, as discussed in the remainder of this chapter.

Table 1
Levels of Evidence for Grading Clinical Utility of Tumor Markers

Level	Type of evidence
I	Evidence from a single, high-powered, prospective, controlled study (with therapy and follow-up dictated by protocol) specifically designed to test marker or evidence from meta-analysis and/or overview of level II or level III studies. Ideally, study is a prospective, controlled randomized trial in which diagnostic and/or therapeutic clinical decisions in one arm are determined at least in part on the basis of marker results, and diagnostic and/or therapeutic clinical decisions in the control arm are made independently of marker results. However, study design may also include prospective but not randomized trials with marker data and clinical outcomes as the primary objective.
II	Evidence from a study in which marker data are determined in relationship to prospective trial that is performed to test therapeutic hypothesis but not specifically designed to test marker utility (i.e., marker study is secondary objective of protocol). However, specimen collection for marker study and statistical analysis are prospectively determined in protocol as secondary objectives.
III	Evidence from large but retrospective studies from which variable numbers of samples are available or selected. Therapeutic aspects and follow-up of patient population may or may not have been prospectively dictated. Statistical analysis for tumor marker was not dictated prospectively at time of therapeutic trial design.
IV	Evidence from small retrospective studies that do not have prospectively dictated therapy, follow-up, specimen selection, or statistical analysis. Study design may use matched case controls, etc.
V	Evidence from small pilot studies designed to determine or estimate distribution of marker levels in sample populations. Study designs may include "correlation" with other known or investigational markers or outcome but is not designed to determine clinical utility.

Adapted from the Tumor Marker Utility Grading System of Hayes et al. *(5)*, with permission from Oxford University Press.

3. Statistical Issues in Prognostic Marker Studies

Statistical issues to be considered in prognostic marker studies are numerous. First, the method in which the marker is acquired may involve laboratory assays and procedures in which reproducibility and validity are concerns. Furthermore, there may be competing laboratory evaluation methods with different scoring systems for a given marker, and various discrete cut-points used for classification of assay results into positive or negative findings. An appropriate evaluation of a new marker must take into account existing prognostic factors, as disease characteristics are often correlated, and a new marker may add little additional information over established factors (which may be easier and more economical to obtain). Thus, modeling with multiple covariates is required, and statistical power for these models may be inadequate, particularly when evaluating whether there exists any differential treatment response associated with marker values, represented by interaction terms in the model. Additional problems include multiplicity issues associated with examining multiple cut-points for a marker and examining multiple related outcome measures. Several excellent summaries of statistical problems in prognostic factor studies have appeared in recent years, and in this chapter we reiterate much of this work (*6–8*). Specific issues related to evaluation of erbB-2 in relation to breast cancer are discussed throughout **Subheading 4**.

3.1. Assay Evaluation

Any laboratory procedure is subject to measurement error, and modern molecular biology techniques in particular may involve complex processes that must be carefully controlled. Validity of results from such assays must be established through standard sensitivity and specificity evaluation, provided that a ''gold standard'' evaluation method and result are available. For new markers and techniques, such a standard may not be available, and expert consensus may be required to standardize and score results. (*See* Chapter 5 for a discussion of many of these issues.) In addition, inter-laboratory variability may need to be accounted for, as many studies in cancer involve the enrollment of patients from multiple institutions where laboratory quality and practice may differ. Finally, most tumor marker studies are conducted retrospectively on archived materials that may be sub-optimal, and it is important to address the validity of findings from such studies in relation to the types of samples that might be used in prospective evaluation of patients in clinical practice.

3.2. Scoring and Classification of Marker Results

The choice of cut-points for discrete classification of assay results is often not well motivated. When multiple classification schemes are investigated, and the grouping that produces the largest difference in outcome subsequently selected, the result can be a serious inflation of the apparent prognostic value of the marker, and other studies may then fail to reproduce the observation (*9,10*). Consequences of evaluating numerous cut-points were illustrated in a study by Hilsenbeck and Clark (*11*). In their study, simulations were conducted whereby multiple cut-points were applied to a continuous null marker (e.g., with no prognostic significance) to create two groups that were then compared in relation to clinical outcomes. Type I error rates increased from the expected 5% to 20–25% and higher when 5–10 candidate cut-points were tested and the maximum test statistic obtained was taken as the overall result of the marker evaluation. The authors also provided a review and comparison of methods for adjustment of *p*-values obtained from testing of prognostic markers. Cut-points might be avoided altogether by using continuous marker values, the functional form of which might be obtained by various exploratory methods such as splines *(12–14)*.

3.3. Statistical Power and Modeling

Statistical power is often inadequate in prognostic factor studies, and because most such studies are retrospective and observational in nature, the problem is further exacerbated by lack of a randomization mechanism, missing or misclassified covariates, and other problems. Several authors have commented on sample size requirements for adequate detection of main and interaction effects in prognostic factor studies. For simplicity of the discussion, we assume here that the marker can be partitioned into a dichotomy. The most common effect measure in prognostic factor studies with survival or related time-to-event endpoints is the risk ratio, which is usually computed from the Cox proportional hazards model *(15)*. No definitive rule exists for effect size, but generally a marker that imparts a risk of twofold or greater would be considered to have clinically consequential potential. For the multiplicative relative risk scale, markers that impart small risks are not likely to be found statistically significant in small samples. The frequency distribution of the marker values will also influence statistical power, and in general these frequencies cannot be manipulated but are subject to the observed prevalence of the marker. In most cases, prevalence of unfavorable values for the marker will not be near the optimal value of 50%. Schoenfeld derived the required sample size for the Cox proportional hazards model, obtaining the same formula as that for two-sample log-rank test comparison under the proportional hazards assumption *(16)*. For a given two-sided significance level α, power $1-\beta$, and risk ratio *(RR)* of interest, the total number of failures required is

$$n = \frac{\left(Z_{1-\alpha/2} + Z_{1-\beta}\right)^2}{\left(\ln(RR)\right)^2 \omega(1-\omega)},$$

where $Z_{1-\alpha/2}$ and $Z_{1-\beta}$ are $100 \times (1 - \alpha/2)\%$ and $100 \times (1 - \beta)\%$ standard normal deviates, respectively, and ω is the proportion of patients with the marker value of interest.

Schmoor and colleagues have extended Schoenfeld's results to account for correlation between the marker of interest and some other covariate, as analysis of new prognostic factors necessitates the consideration of known prognostic markers *(17)*. Their derivation results in a straightforward modification of the above equation that incorporates a "variance inflation factor" to account for correlation between model covariate X_1 and another covariate (or composite of other covariates) X_2:

$$n = \frac{(Z_{1-\alpha/2} + Z_{1-\beta})^2}{(\ln(RR))^2 \omega(1-\omega)} \cdot \left(\frac{1}{1-\rho^2}\right),$$

where ρ is the correlation between X_1 and X_2.

For interaction effects, the situation is more challenging. Petersen and George *(18)* addressed sample size requirements for study designs of interaction effects in $2 \times K$ factorial experiments, where there are two treatments and a marker takes on $k = 2, 3,.., K$ values. Again, a modification of the usual sample size formula for hazard ratios is obtained. For the case of two treatment groups ($i = 1, 2$) and a two level prognostic marker ($j = 1, 2$), we define $\Delta_1 = \lambda_{11}/\lambda_{21}$, the treatment hazard ratio for level 1 of the marker and $\Delta_2 = \lambda_{12}/\lambda_{22}$, the treatment hazard ratio for level 2 of the marker. We wish to test H_o: $\Delta_1/\Delta_2 = 1.0$ using a two-sided α level test with power $1 - \beta$ against a specific interaction effect $\Delta_1/\Delta_2 = \theta \neq 1.0$. Under a proportional hazards assumption, the estimator $\ln(\Delta_1/\Delta_2)$ has variance approximately equal to $\sum_{ij} 1/n_{ij}$, where n_{ij} is the number of failures observed in treatment i and marker level j. It follows that the number of failures needed to achieve power $1 - \beta$ must approximately satisfy

$$\left(\sum_{ij} 1/n_{ij}\right)^{-1} = \frac{(Z_{1-\alpha/2} + Z_{1-\beta})^2}{(\ln(\Delta_1/\Delta_2))^2}.$$

Using the fact that the harmonic mean of the n_{ij}'s is less than or equal to the arithmetic mean, this equation shows that the total number of failures required to detect a treatment by marker interaction with power of $1 - \beta$ is at least four times greater than the number of failures needed to detect a similarly sized treatment hazard ratio within a population that is homogeneous with respect to the prognostic marker. In designed experiments where treatment allocation could be balanced within strata of marker values via prospective sampling (so that $n_{1j} \approx n_{2j}$ for $j = 1, 2$), then the "four times greater" rule holds well; in cases of unequal frequencies of n_{ij}, sample size requirements will be even larger. Schmoor and colleagues also addressed interaction effects taking other covariates into consideration for the case of exponential failure times, which provides an approximate solution for the more general case *(17)*.

In addition to statistical power, there are numerous other statistical considerations in prognostic marker studies specifically related to modeling. These include verification of the correct model form, variable selection methods (e.g., stepwise regression and others), the aforementioned issues concerning definitions of discrete covariates, and model validation on independent data. A detailed discussion of concerns related to modeling can be found in Simon and Altman *(6)* and George *(7)*. Klinger and colleagues discuss some alternatives to typical survival analysis modeling methods, such as regression trees, in the context of oncology research *(19)*.

4. Case Study: *erbB-2* and Breast Cancer Treatment Response
4.1. Background: erbB-2 *and Breast Cancer*

The *erbB-2* oncogene (also known as c-*erbB-2* and HER2/*neu*), which encodes a specific transmembrane growth factor receptor of the tyrosine kinase family, was found to be amplified in a human breast carcinoma cell line by King and colleagues in 1985 *(20)*. Subsequently, Slamon and colleagues reported that amplification of the *erbB-2* gene was present in 20–30% of breast cancers and was associated with shorter survival and disease-free survival time *(21,22)*. It was conjectured that the basis for this association was a greater cell proliferation rate in tumors with erbB-2 amplification. These and subsequent analyses showed that erbB-2 (either amplification of the gene or overexpression of its protein product) was prognostic among both patients with tumors that had spread to the axillary lymph nodes (node-positive patients) and among patients with tumors confined to the breast (node-negative patients) *(23–25)*. Other reports, however, did not confirm the relationship, or did not show a strong independent prognostic value for erbB-2, and there has been controversy in establishing the role of the marker as a clinically useful prognostic factor *(26–29)*. Some authors have related this controversy directly to issues concerning laboratory evaluation of the marker *(23,30–32)*. Nevertheless, the weight of evidence currently suggests that overexpression of erbB-2 does impart a less favorable prognosis. A recent meta-analysis of approx 35 studies appearing between 1996 and 1999 found erbB-2 to be a moderate but not particularly strong risk factor for breast cancer recurrence and death *(33)*.

Early studies of erbB-2 suggested that it was not only associated with poor prognosis but also with a differential benefit depending on the chemotherapy drug or regimen administered. Several studies suggested that tumors with overexpression did not respond as well to cyclosphosphamide, methotrexate, and fluorouracil (CMF, a commonly used chemotherapy regimen) as erbB-2 negative tumors *(26,27,34)*, while others did not confirm this finding *(35,36)*. Other studies suggested that erbB-2 overexpressing tumors were less sensitive to tamoxifen *(37)*, again an observation not confirmed by others *(38,39)*.

Overexpression of erbB-2 was more convincingly correlated with response to regimens containing doxorubicin (commercially, Adriamycin, a member of a class of agents known as anthracyclines) in a series of studies appearing in the middle to late 1990s. Muss and colleagues first reported that a more intensive dose of cyclophosphamide, doxorubicin, and fluorouracil was of greater benefit among erbB-2 positive patients than among those with tumors that did not overexpress erbB-2 (*40*). A subsequent analysis of the same patient cohort and additional data also supported this conclusion (*41*). Independent reports, one of which is discussed in **Subheading 4.2.**, have also supported an association between overexpression and response to doxorubicin, and thus suggest that one might use the marker in clinical practice to choose treatment, as least for this agent (*42,43*). Additional investigations have explored whether those with overexpression would preferentially benefit from taxanes, but little reliable information is available thus far. Finally, a targeted agent for the erbB-2 receptor, trastuzumab (Herceptin, commercially), has appeared to show preferential efficacy among tumor cells overexpressing erbB-2 in preclinical studies (*44*). Thus far, efficacy trials of trastuzumab in humans have been conducted exclusively among erbB-2 positive patients. Assuming the mechanism of action is correct, it is plausible that little or no benefit would be realized for this agent among those whose tumors are erbB-2 negative.

4.2. The National Surgical Adjuvant Breast and Bowel Project B-11 Trial

The National Surgical Adjuvant Breast and Bowel Project (NSABP) is a federally funded multicenter cooperative clinical trials group that has carried out studies addressing the treatment and prevention of breast and colorectal cancers. A spectrum of modalities has been investigated, including surgical procedures, radiotherapy, chemotherapy, hormonal therapy, and biologic agents. In parallel with this effort, pathologic materials are collected and analyzed to investigate ancillary questions in the natural history and treatment of these cancers. Pathology materials are evaluated concurrently with conduct of the studies, and are also archived for future use.

In an earlier NSABP study, erbB-2 protein overexpression was found to be associated with poorer survival prognosis and other unfavorable pathologic features among patients with either node-negative or node-positive breast cancer (*45*). Subsequently, the potential for differential response to therapy according to erbB-2 status was investigated in NSABP protocol B-11, a randomized clinical trial evaluating the addition of doxorubicin to a two-drug chemotherapy regimen of L-phenylalanine mustard and fluorouracil (denoted PF) (*43,46*). Although PF has been superseded as a treatment option for breast cancer and other trials were available for erbB-2 evaluation, because the B-11 regimens differed only by the addition of doxorubicin, the trial was selected for evaluation first as a "proof of principle" study to address the potential erbB-2–doxorubicin interaction. What follows is a detailed description of the analysis.

In protocol B-11, women with lymph node positive operable breast cancer were treated by either radical or modified radical mastectomy and randomized to receive either (1) PF or (2) PF and doxorubicin (PAF). Between June 1981 and September 1984, 707 patients were randomized, of whom 682 met study eligibility requirements. Further details of the study design and primary findings have been published previously (46). Endpoints for evaluation of erbB-2 in relation to response to doxorubicin were the same as those for the primary analysis of B-11. Disease-free survival (DFS) time was defined as time from surgery until breast cancer recurrence at any local, regional, or distant anatomic site, new primary cancer of any site, or death prior to these events. Survival time was defined as time until death from any cause. Two additional secondary endpoints, distant disease-free survival (DDFS) and recurrence-free survival (RFS), were addressed in the study but are not presented here.

4.2.1. Evaluation of erbB-2 in NSABP Protocol B-11 Tumor Samples

While about 200 patients had paraffin-embedded tumor blocks, > 90% (638 patients) had archived precut unstained tumor sections or hematoxylin–eosin (H&E) stained sections prepared as slides, and, consequently, these materials were used to perform evaluation for erbB-2. Such material is amenable to immunohistochemical (IHC) analysis to determine erbB-2 protein overexpression. IHC staining was performed using a cocktail of two antibodies (described in detail in [43]) using both the unstained materials and stained slides. The determination of whether there was overexpression was based on a simple dichotomy, whereby the reaction was scored as positive if any cells showed definitive staining. Two individuals rated the slides together while blinded as to treatment assignment or outcome of the patient.

In this analysis, one immediate concern was whether the unstained and stained slides could be similarly stained and interpreted for erbB-2 and thus a simple sensitivity and specificity analysis was conducted. A comparison of staining sensitivity was performed whereby the assays of 51 cases were replicated using both stained and unstained sections available on the same patients. A simple cross-tabulation of positive and negative findings according to the two methods indicated 98% agreement (50 of 51 cases). Another quality assessment of materials involved a comparison of freshly cut sections from the paraffin blocks and previously cut and prepared slides. Sixty cases for which paraffin blocks and slides were available were assessed, and a 12% false-negative rate (25 were positive in fresh section, 22 were positive in slides) was observed. Thus, an analysis based on slide materials may be biased toward an attenuation of the effect of erbB-2 positivity, in that erbB-2 positive cases may be classified as negative.

Scoring methods for IHC and other assays of erbB-2 have been the subject of considerable controversy (47). For the IHC analysis results in the NSABP study, the percentage of positive staining for each patient's specimen was computed. The distribution was highly bimodal at 0% and 100%, suggesting that the dichotomous classification was the best approach with the available material. A further rationale for choosing the dichotomous rating system was that, in unstained and H&E stained slides rather than fresh material, it was deemed difficult to ascribe meaning to the quantitative percentage of cells staining, as has been done by other investigators, because the result could be largely an artifact of the laboratory procedure. Questions of assay reliability for IHC methods have led some to suggest that fluorescence *in situ* hybridization (FISH) analysis, which measures gene amplification (copy number) rather than protein expression, would be preferable (24,30,48).

4.2.2. Relationship of erbB-2 to Other Patient and Tumor Characteristics

Because negative and positive prognostic factors are often interrelated, the joint distribution of erbB-2 overexpression and other patient and tumor characteristics was examined. About 37.5% of patients exhibited erbB-2 positive tumors. Examining the cross-classification of factors singly with erbB-2, it was found that a higher number of positive nodes, larger tumor size, and estrogen receptor negative tumors were associated with erbB-2 overexpression. To take factors into account jointly, a logistic regression model relating erbB-2 to all covariates was employed, yielding similar results.

4.2.3. erbB-2 as a Predictor of Response to Doxorubicin

The explicit hypothesis of this investigation was that the benefit of doxorubicin would be largely confined to those patients with erbB-2 overexpression, that is, outcomes would differ in favor of PAF among patients with overexpression, while among those without overexpression, outcomes for PF and PAF would be similar. Accordingly, comparisons of treatment outcomes were conducted separately for the cohorts of erbB-2 negative (n =399) and erbB-2 positive (n = 239) patients. Kaplan–Meier estimates of disease-free survival (DFS) and survival are shown in **Fig. 1**. For each erbB-2 cohort, a PAF/PF relative risk (RR) estimate and corresponding significance test for the null hypothesis $RR = 1.0$ were obtained by the Cox proportional hazards model containing other relevant prognostic covariates (patient age at surgery, clinical tumor size, lymph node status, and estrogen receptor status). Results suggested that the benefit of doxorubicin (e.g., the PAF treatment arm) was evident only for those patients overexpressing erbB-2. This was confirmed by a formal test of differential benefit for doxorubicin according to erbB-2 status by combining all patients and testing an interaction term in the proportional hazards model (**Fig. 2**). The resulting interaction tests for the various endpoints were statisti-

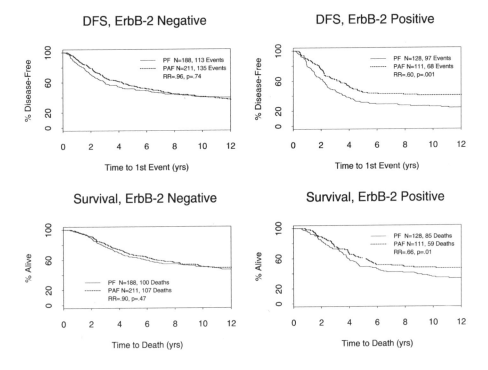

Fig. 1. Kaplan–Meier plots for PF and PAF treatment arms according to erbB-2 status. Two endpoints, disease-free survival (DFS) *(top row)* and survival *(bottom row)*, are shown. Relative risks *(RR)* and *p*-values shown on each plot are from the Cox proportional hazards model with covariates for treatment, age at surgery, clinical tumor size, pathologic lymph node status, and estrogen receptor expression.

cally significant or nearly significant at conventional levels. Similar results were obtained for the RFS and DDFS endpoints.

In the B-11 study, the DFS and survival endpoints were considered primary and the other endpoints secondary, and findings for all four endpoints were consistent. Nevertheless, concern over multiplicity of hypothesis tests prompted the determination of a *p*-value for the interaction effect adjusted for the number of tests. A Bonferroni type adjustment, whereby one multiplies the minimum *p*-value by 4, would constitute an overly conservative adjustment here, because test statistics for the four endpoints are highly correlated. Instead, bootstrap resampling was used to estimate the correlation among the four tests, and the *p*-value associated with the maximum absolute Z value was computed by numerical integration. The adjusted *p*-value obtained for the hypothesis of interaction between treatment and erbB-2 status was 0.04.

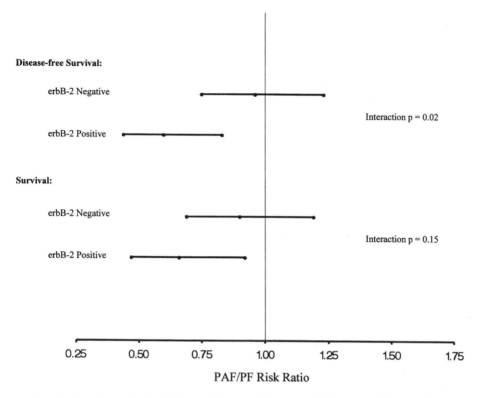

Fig. 2. Relative risk for PAF vs. PF according to erbB-2 status. The *p*-values for interaction between treatment and erbB-2 status are from a Wald test of the cross-product term of these covariates in the Cox proportional hazards model.

4.3. A Prospective Study for erbB-2 Targeted Therapy

As a targeted therapeutic strategy, researchers have created a monoclonal antibody to bind to the extracellular domain of the growth factor receptor encoded by the *erbB-2* oncogene, with the goal of inhibiting growth of tumors overexpressing the receptor. Through cell line experiments and animal xenograft models, the antibody was demonstrated to have significant antiproliferative effects. A genetically engineered successor agent (trastuzumab, Herceptin, commercially) was subsequently developed and found to be efficacious in patients with metastatic breast cancer overexpressing erbB-2 *(49,50)*. The NSABP as well as other clinical trials groups have recently begun trials to evaluate Herceptin in the adjuvant setting. The NSABP trial (protocol B-31) will compare doxorubicin and cyclophosphamide followed by taxol with that same regimen plus Herceptin, in patients with operable erbB-2 positive tumors and positive lymph nodes. This study will enroll 2700 patients and require roughly 8 yr until definitive results are obtained. Several ancillary investigations concerning erbB-2 are planned, as described in the following paragraphs.

In this multiinstitutional study, participating centers will perform erbB-2 testing using either IHC or FISH and score results as positive or negative according to a common criterion (only patients testing positive will be enrolled). Tumor specimens will also be provided to the central pathology laboratory, where a comprehensive reevaluation of erbB-2 will be conducted in order to address several ancillary study aims. These are: (1) to verify the reported status of the tumor from the institution; (2) to compare results of the various assays, with the opportunity for a direct comparison of methods that measure protein overexpression and those which evaluate gene amplification; and (3) to evaluate whether the assays can predict response to Herceptin. Six assay types will be performed: the DAKO HercepTest kit, TAB250, TAB250/ pAb-1 cocktail (used in the B-11 analysis), CB-11, HER-2 FISH assay, and array-based CGH. Several other important pathologic studies will also be performed. One of these involves determining whether expression of the phosphorylated erbB-2 receptor (an indicator that the receptor is capable of binding) in the tumor is prognostic for outcomes or predictive of response to Herceptin, and to evaluate whether this frequency of expression differs in postrelapse tissue among patients either receiving or not receiving Herceptin. Another involves evaluating whether shed extracellular domain of erbB-2 or autoantibodies found in patient serum are associated with outcomes or response to Herceptin.

Explicit hypotheses and analytic methods for these investigations are described in the B-31 protocol document. By the addition of these ancillary studies to this randomized trial designed to evaluate the efficacy of Herceptin in addition to multidrug chemotherapy among erbB-2 positive patients, the B-31 study will address in a comprehensive and prospective manner many of the outstanding research questions concerning erbB-2.

5. Summary and Recommendations for Analyzing Molecular Markers in Clinical Cancer Research

The case study presented here illustrates how previously collected clinical outcomes data, even with a "retired" treatment regimen, can serve as a vital resource for advancing the understanding of the natural history of cancer and, furthermore, can play a role in refining treatment selection for current patients. The availability of archived tumor samples allowed for the augmentation of the long-term outcome data under randomized treatment assignment with a modern molecular marker and yielded an important finding of biologic and clinical relevance today. This study, combined with the foundational work that preceded it and the concurrent results from other large clinical trials groups, have fostered debate regarding breast cancer clinical practice, with erbB-2 evaluation now being advocated by some experts (but not others) as part of a routine clinical evaluation prior to treatment *(28,51)*. In a recent review and consensus statement regarding prognostic factors in breast cancer from the College of American Pathologists *(52)*, a detailed list of issues, guidelines, and recommendations related to erbB-2 were provided, many of which were reviewed in this discussion. It is clear that erbB-2 will remain an important research area in breast cancer treatment.

Despite the apparent interaction between erbB-2 and response to doxorubicin-containing chemotherapy regimens, it has yet to be established unequivocally which mechanism is at play in this relationship. In the NSABP study, it can be argued that two factors varied between the PF and PAF groups: the addition of doxorubicin and a simple increase in total chemotherapy dose via the use of three agents instead of two. Similarly, in the CALGB study, the dose of doxorubicin and the other agents varied between treatment groups. Thus, in either study, it can be conjectured that (1) erbB-2 positive tumors may be specifically sensitive to doxorubicin or (2) erbB-2 tumors may be more resistant to chemotherapy and that greater total chemotherapy exposure is beneficial. There is supporting biological information for both mechanisms. Statisticians and other researchers should take heed that, particularly in retrospective studies, observations may be simultaneously consistent with several hypotheses.

To this end, a follow-up investigation by the NSABP has explored further the hypothesis that doxorubicin-containing chemotherapy regimens might specifically be more advantageous in patients with erbB-2 overexpression. Among 2295 eligible node-positive patients entered onto NSABP Protocol B-15, a randomized trial comparing AC, CMF, and a regimen of AC followed by reinduction CMF *(53)*, 2034 (89%) had immunohistochemical analysis of erbB-2. Statistical analyses were similar to that of B-11, with the primary study hypothesis being whether there was a differential benefit from the doxorubicin-containing regimen (AC) relative to CMF according to erbB-2 status. Findings indicated that the superiority of AC over CMF was restricted to erbB-2 positive patients, although the differences did not reach statistical significance *(54)*. These results provide further evidence that a regimen containing doxorubicin (or other anthracyclines) is preferred for patients with erbB-2 positive tumors, and unlike the B-11 study, directly addresses current treatment guidelines, as AC and CMF remain in wide use.

In this chapter we have described a retrospective analysis where current clinical data and archived biologic samples were used to address a current question in breast cancer. Despite limitations of the materials and the retrospective nature of the investigation, a pragmatic and thoughtful analysis yielded valuable information. With the proliferation of biotechnology, there is an ever-greater need to evaluate markers in cancer. In recent years, statisticians have provided extensive comment on the past and current state of research in marker studies, and have proposed appropriate prospective study designs to improve the quality of research. For example, Simon and Altman describe a study classification scheme similar to that used to describe studies in the evolution of therapeutic agents *(6)*. Phase I studies are those preliminary studies that establish the potential worth of a given marker. Phase II studies are small, exploratory studies that often demonstrate the usefulness of a marker under less than ideal circumstances. Phase III prognostic factor studies are large, definitive studies that provide a considerable weight of evidence for or against a marker's clinical utility. Their criteria for phase III prognostic factor studies are as follows: First, a valid, reproducible assay will be needed, with documentation of inter- and intra-laboratory variation. Assessors of assay results should be blinded to clinical outcome. The study group should be a well-defined cohort for whom the study referral pattern and eligibility are described. The number of patients subsequently unevaluable for the marker should be small (preferably <15%). Treatment should be standardized for all patients or be randomized. Hypotheses should be stated a priori and include specification of endpoints, definition of scoring for the marker result, identification of patient subsets of interest, and other prognostic factors to be included in the analysis. The number of patients and events should be sufficiently large so that power is adequate for clinically relevant effects. Multiple regression models or stratification methods should be used to establish that the marker is prognostic over and above other known prognostic factors. Confidence limits should be presented for effect measures, with multiplicity of tests taken into consideration.

Statisticians contributing to the summary statement on prognostic factors from the College of American Pathologists *(4)* similarly presented a broad and comprehensive view of the design needs of future prognostic factor studies (**Table 2**). These considerations, as well as the comments and recommendations of authors cited throughout this chapter, should serve as a valuable guide for the applied statistician engaged in this important research area.

Acknowledgments

This work was supported by Public Health Service Grants U10-CA-76001 (Dignam), U10-CA-12027 (Paik), and U10-CA-69651 (Dignam, Bryant) from the National Cancer Institute, National Institutes of Health, Department of Health and Human Services.

Table 2
General Statistical Recommendations from
College of American Pathologists Conference XXXV

1. Clinical trials should be specifically designed to test whether a factor has prognostic value. This question can be included in a therapeutic trial, but careful attention must be paid that there is sufficient statistical power to answer both the prognostic and the treatment questions.

2. Prognostic factor question must be prioritized for importance by multidisciplinary groups of investigators working with each cancer type so that the most important factors are quickly evaluated.

3. Journals should adopt publication guidelines for reporting results from prognostic factor studies, including the following elements:

    a. Assessment of possible patient selection bias

        i. Source of patients for the study
        ii. Difference between patients with and without tumor marker in terms of

            1. Baseline demographic and tumor characteristics
            2. Treatment received
            3. Efficacy outcomes

    b. Statement about how missing data were handled
    c. Cut-point selection for method stated
    d. Adjustments for multiple testing
    e. Statistical power analysis if conclusion is negative
    f. Large validation studies should be given publication preference after initial exploratory work for a tumor marker.

4. Organization addressing prognostic factor categorization should come to consensus about the ranking of factors, or at least harmonize their recommendations relative to each other, so that a clear picture of the relative value of various factors is developed.

5. Continued research into multivariate analysis techniques for incomplete data and for evaluation of multiple factors is needed.

Reprinted from **ref.** *4* with permission from the College of American Pathologists.

References

1. McGuire, W. L. and Clark, G. M. (1992) Prognostic factors and treatment decisions in axillary-node-negative breast cancer. *N. Engl. J. Med.* **326,** 1756–1761.
2. Goldhirsch, A., Glick, J. H., Gelber, R. D., and Senn, H. J. (1998) Meeting highlights: International Consensus Panel on the Treatment of Primary Breast Cancer. *J. Natl. Cancer Inst.* **90,** 1601–1608.

3. Thomssen, C., Janicke, F., Kaufmann, M., Scharl, A., and Hayes, D. F. (2000) Do we need better prognostic factors in node-negative breast cancer? *Eur. J. Cancer* **36**, 293–306.
4. Hammond, M. E., Fitzgibbons, P. L., Compton, C. C., Grignon, D. J., Page, D. L., Fielding, L. P., et al. (2000) College of American Pathologists Conference XXXV: Solid tumor prognostic factors—which, how, and so what? *Arch. Pathol. Lab. Med.* **124**, 958–965.
5. Hayes, D. F., Bast, R. C., Desch, C. E., Fritsche, H., Jr, Kemeny, N. E., Jessup, J. M., et al. (1996) Tumor marker utility grading system (TMUGS): a framework to evaluate clinical utility of tumor markers. *J. Natl. Cancer Inst.* **88**, 1456–1466.
6. Simon, R. and Altman, D. G. (1994) Statistical aspects of prognostic factor studies in oncology. *Br. J. Cancer* **69**, 979–985.
7. George, S. L. (1994) Statistical considerations and modeling of clinical utility of tumor markers, in *Hematology/Oncology Clinics of North America: Tumor Markers in Adult Solid Malignancies* (Hayes, D. F., ed.) Saunders, Philadelphia, PA, pp. 457–470.
8. Pajak, T. F., Clark, G. M., Sargent, D. J., McShane, L. M., and Hammond, E. H. (2000) Statistical issues in tumor marker studies. *Arch. Pathol. Lab. Med.* **124**, 1011–1015.
9. Hilsenbeck, S. G., Clark, G. M., and McGuire, W. L. (1992) Why do so many prognostic factors fail to pan out? *Breast Cancer Res. Treat.* **22**, 197–206.
10. Altman, D. G., Lausen, B., Sauerbrei, W., and Schumacher, M. (1994) Dangers of using "optimal" cutpoints in the evaluation of prognostic factors. *J. Natl. Cancer Inst.* **86**, 829–835.
11. Hilsenbeck, S. G. and Clark, G. M. (1996) Practical p-value adjustment for optimally selected cutpoints. *Statist. Med.* **15**, 103–112.
12. Durrleman, S. and Simon, R. (1988) Flexible regression models with cubic splines. *Statist. Med.* **8**, 551–561.
13. Gray, R. (1992) Flexible methods for analyzing survival data using splines, with applications to breast cancer prognosis. *J. Am. Statist. Assoc.* **87**, 942–951.
14. Bryant, J., Fisher, B., Gunduz, N., Costantino, J. P., and Emir, B. (1998) S-phase fraction combined with other patient and tumor characteristics for the prognosis of node-negative, estrogen-receptor-positive breast cancer. *Breast Cancer Res. Treat.* **51**, 239–253.
15. Cox, D. R. (1972) Regression models and life table. *J. Roy Statist. Soc.* B **34**, 187–220.
16. Schoenfeld, D. A. (1983) Sample-size formula for the proportional-hazards regression model. *Biometrics* **39**, 499–503.
17. Schmoor, C., Sauerbrei, W., and Schumacher, M. (2000) Sample size considerations for the evaluation of prognostic factors in survival analysis. *Statist. Med.* **19**, 441–452.
18. Peterson, B. and George, S. L. (1993) Sample size requirements and length of study for testing interaction in a 2×k factorial design when time-to-failure is the outcome. *Control Clin. Trials* **14**, 511–522. (Erratum [1994] *Control Clin. Trials* **15**, 326).

19. Klinger, A., Donnegger, F., and Ulm, K. (2000) Identifying and modeling prognostic factors with censored data. *Statist. Med.* **19**, 601–615.

20. King, C. R., Kraus, M. H., and Aaronson, S. A. (1985) Amplification of a novel v-*erbB*-related gene in a human mammary carcinoma. *Science* **229**, 974–976.

21. Slamon, D. J., Clark, G. M., Wong, S. G., Levin, W. J., Ullrich, A., and McGuire, W. L. (1987) Human breast cancer: correlation of relapse and survival with amplification of HER-2/*neu* oncogene. *Science* **235**, 177–182.

22. Slamon, D. J., Godolphin, W., Jones, L. A., Holt, J. A., Wong, S. G., Keith, D. E., et al. (1989) Studies of HER2/neu proto-oncogene in human breast and ovarian cancer. *Science* **244**, 707–712.

23. Press, M. F., Pike, M. C., Chazin, V. R., Hung, G., Udove, J. A., Markowicz, M., et al. (1993) HER-2/neu expression in node-negative breast cancer: direct tissue quantitation by computerized image analysis and association of overexpression with increased risk of recurrent disease. *Cancer Res.* **53**, 4960–4970.

24. Press, M. F., Bernstein, L., Thomas, P. A., Meisner, L. F., Zhou, J. Y., Ma, Y., et al. (1997) HER-2/*neu* gene amplification characterized by fluorescence in situ hybridization: poor prognosis in node-negative breast carcinomas. *J. Clin. Oncol.* **15**, 2894–2904.

25. Andrulis, I. L., Bull, S. B., Blackstein, M. E., Sutherland, D., Mak, C., Sidlofsky, S., et al. (1998) neu/erbB-2 amplification identifies a poor-prognosis group of women with node-negative breast cancer. *J. Clin. Oncol.* **16**, 1340–1349.

26. Allred, D. C., Clark, G. M., Tandon, A. K., Molina, R., Tormey, D. C., Osborne, C. K., et al. (1992) HER-2/*neu* in node-negative breast cancer: prognostic significance of overexpression influenced by the presence of in situ carcinoma. *J. Clin. Oncol.* **10**, 599–605.

27. Gusterson, B. A., Gelber, R. D., Goldhirsch, A., Price, K. N., Save-Soderborgh, J., Anbazhagan, R., et al. (1992) Prognostic importance of c-erbB-2 expression in breast cancer. International (Ludwig) Breast Cancer Study Group. *J. Clin. Oncol.* **10**, 1049–1056.

28. Clahsen, P. C., van de Velde, C. J., Duval, C., Pallud, C., Mandard, A. M., Delobelle-Deroide, A., et al. (1998) p53 protein accumulation and response to adjuvant chemotherapy in premenopausal women with node-negative early breast cancer. *J. Clin. Oncol.* **16**, 470–479.

29. Piccart, M. J., Di Leo, A., and Hamilton, A. (2000) HER2. a 'predictive factor' ready to use in the daily management of breast cancer patients? *Eur. J. Cancer* **36**, 1755–1761.

30. Press, M. F., Hung, G., Godolphin, W., and Slamon, D. J. (1994) Sensitivity of HER2/neu antibodies in archived tissue samples: potential sources of error immunohistochemical studies of oncogene expression. *Cancer Res.* **54**, 2771–2777.

31. Clark, G. M. (1998) Should selection of adjuvant chemotherapy for patients with breast cancer be based on erbB-2 status? *J. Natl. Cancer Inst.* **90**, 1320–1321.

32. Ravdin, P. M. (1999) Should HER2 status be routinely measured for all breast cancer patients? *Semin. Oncol.* **26**, 117–123.

33. Trock, B. J., Yamauchi, H., Brotzman, M., Stearns, V., and Hayes, D. F. (2000) C-erbB2 as a prognostic factor in breast cancer (BC): a meta-analysis. *Proc. Am. Soc. Clin. Oncol.* **19,** 97.

34. Stal, O., Sullivan, S., Wingren, S., Skoog, L., Rutqvist, L. E., Carstensen, J. M., and Nordenskjold, B. (1995) c-erbB-2 expression and benefit from adjuvant chemotherapy and radiotherapy of breast cancer. *Eur. J. Cancer* **31A,** 2185–2190.

35. Miles, D. W., Harris, W. H., Gillett, C. E., Smith, P., and Barnes, D. M. (1999) Effect of c-erbB(2) and estrogen receptor status on survival of women with primary breast cancer treated with adjuvant cyclophosphamide/methotrexate/fluorouracil. *Int. J. Cancer* **84,** 354–359.

36. Ménard, S., Valagussa, P., Pilotti, S., Biganzoli, E., Boracchi, P., Casalini, P., et al. (1999) Benefit of CMF Treatment in Lymph Node-Positive Breast Cancer Overexpressing HER2. *Proc. Am. Soc. Clin. Oncol.* **17,** 257.

37. Bianco A. R., De Laurentiis, M., Carlomagno, C., Lauria, R., Petrella, G., Panico, L., et al. (1998) 20 year update of Naples GUN trial of adjuvant breast cancer therapy: evidence of interaction between c-erbB-2 expression and tamoxifen efficacy. *Proc. Am. Soc. Clin. Oncol.* **17,** 373.

38. Elledge, R. M., Green, S., Ciocca, D., Pugh, R., Allred, D. C., Clark, G. M., et al. (1998) HER-2 expression and response to tamoxifen in estrogen receptor-negative breast cancer: a Southwest Oncology Group study. *Clin. Cancer Res.* **4,** 7–12.

39. Berry, D. A., Muss, H. B., Thor, A. D., Dressler, L., Liu, E. T., Broadwater, G., et al. (2000) HER-2/neu and p53 expression versus tamoxifen resistance in estrogen receptor–positive, node-positive breast cancer. *J. Clin. Oncol.* **18,** 3471–3479.

40. Muss, H. B., Thor, A. D., Berry, D. A., Kute, T., Liu, E. T., Koerner, F., et al. (1994) c-erbB-2 expression and response to adjuvant therapy in women with node-positive early breast cancer. *N. Engl. J. Med.* **330,** 1260–1266. (Erratum [1994] *N. Engl. J. Med.* **331,** 211).

41. Thor, A. D., Berry, D. A., Budman, D. R., Muss, H. B., Kute, T., Henderson, I. C., et al. (1998) erbB-2, p53, and the efficacy of adjuvant therapy in lymph node-positive, hormone receptor-negative breast cancer. *J. Natl. Cancer Inst.* **90,** 1346–1360.

42. Ravdin, P. M., Green, S., Albain, K. S., Boucher, V., Ingle, J., Pritchard, K., et al. (1998) Initial report of the SWOG biologic correlation study of c-erbB-2 expression as a predictor of outcome in a trial comparing adjuvant CAFT to tamoxifen alone. *Proc. Am. Soc. Clin. Oncol.* **17,** 374.

43. Paik, S., Bryant, J., Park, C., Fisher, B., Tan-Chiu, E., Hyams, D., et al. (1998) erbB-2 and response to doxorubicin in patients with axillary lymph node-positive, hormone receptor-negative breast cancer. *J. Natl. Cancer Inst.* **90,** 1361–1370.

44. Pietras, R. J., Fendly, B. M., Chazin, V. R., Pegram, M. D., Howell, S. B., and Slamon, D. J. (1994) Antibody to HER-2/neu receptor blocks DNA repair after cisplatin in human breast and ovarian cancer cells. *Oncogene* **9,** 1829–1838.

45. Paik, S., Hazan, R., Fisher, E. R., Sass, R. E., Fisher, B., Redmond, C., et al. (1990) Pathologic findings from the National Surgical Adjuvant Breast and Bowel Project: prognostic significance of erbB-2 protein overexpression in primary breast cancer. *J. Clin. Oncol.* **81,** 103–112.

46. Fisher, B., Redmond, C., Wickerham, D. L., Bowman, D., Schipper, H., Wolmark, N., et al. (1989) Doxorubicin containing regimens for the treatment of stage II breast cancer: the National Surgical Breast and Bowel Project experience. *J. Clin. Oncol.* **7,** 572–582.
47. Nelson, N. J. (2000) Experts debate value of HER2 testing methods [News]. *J. Natl. Cancer Inst.* **92,** 292–294.
48. Mitchell, M. S. and Press, M. F. (1999) The role of immunohistochemistry and fluorescence in situ hybridization for HER2/neu in assessing the prognosis of breast cancer. *Semin. Oncol.* **26,** 108–116.
49. Cobleigh, M. A., Vogel, C. L., Tripathy, D., Robert, N. J., Scholl, S., Fehrenbacher, L., et al. (1999) Multinational study of the efficacy and safety of humanized anti-HER2 monoclonal antibody in women who have HER2-overexpressing metastatic breast cancer that has progressed after chemotherapy for metastatic disease. *J. Clin. Oncol.* **17,** 2639–2648.
50. Norton, L., Slamon, D., Leyland-Jones, B., Wolter, J., Fleming, T., Eirmann, W., et al. (1999) Overall survival (os) advantage to simultaneous chemotherapy (CRx) plus the humanized anti-HER2 monoclonal antibody Herceptin (H) in HER2-overexpressing (HER2+) metastatic breast cancer (MBC). *Proc. Am. Soc. Clin. Oncol.* **18,** 483.
51. Hayes, D. F., Yamauchi, H., Stearns, V., Brotzman, M., Isaacs, C., and Trock B. (2000) Should all breast cancers be tested for c-erbB-2? in *2000 ASCO Educational Book* (Perry, M. C., ed.). American Society of Clinical Oncology, Alexandria, VA, pp. 257–265.
52. Fitzgibbons, P. L., Page, D. L., Weaver, D., Thor, A. D., Allred, D. C., Clark, G. M., et al. (2000) Prognostic factors in breast cancer: College of American Pathologists Consensus Statement 1999. *Arch. Pathol. Lab. Med.* **124,** 966–978.
53. Fisher, B., Brown, A. M., Dimotrov, N. V., Poisson, R., Redmond, C., Margolese, R. G., et al. (1990) Two months of doxorubicin-cyclophosphamide with and without interval reinduction therapy compared with 6 months of cyclophosphamide, methotrexate, and fluorouracil in positive-node breast cancer patients with tamoxifen-nonresponsive tumors: results from the National Surgical Adjuvant Breast and Bowel Project B-15. *J. Clin. Oncol.* **8,** 1483–1496.
54. Paik, S., Bryant, J., Tan-Chiu, E., Yothers, G., Park, C., Wickerham, D. L., and Wolmark, N. (2000) HER2 and choice of adjuvant chemotherapy for invasive breast cancer: NSABP Protocol B-15. *J. Natl. Cancer Inst.* **92,** 1991–1998.

10

Power of the Rank Test for Multi-Strata Case-Control Studies with Ordinal Exposure Variables

Grzegorz A. Rempala and Stephen W. Looney

1. Introduction

In epidemiological studies of rare diseases (i.e., rare types of cancer) researchers often face major difficulty in obtaining enough cases of the disease to make valid comparisons using odds-ratio estimators. Moreover, they may wish to adjust for the influence of certain extraneous factors so that the effect of the variables of interest can be more clearly visible. This is especially so in case-control studies when it is known that the effects of the risk factor are confounded with such variables as age, sex, and individual physical characteristics of the subjects. These confounding variables often make it difficult (or even impossible) to directly compare the exposed and unexposed groups. Typically, to evaluate the effect of the risk factor in these situations within the odds-ratio framework, methods based on data stratification and within-stratum dichotomization are used. The latter is usually accomplished by classifying cases and controls within each stratum as either exposed or unexposed to the risk factor under investigation. Whereas the stratification is often unavoidable, it may not be practical to dichotomize exposure. Instead, one might wish to consider multiple levels of the exposure variable, based on some appropriate ordinal or even continuous scale (cf., e.g., Greenberg and Tamburro *[1]*).

The statistical problem of testing for the exposure effect in such settings has been considered by several authors and a variety of approaches have been discussed. In particular, a test based on the rank of the exposure level of each case within a group of individually matched controls was proposed *(2)* and a study of its large sample properties under the null hypothesis followed *(3)*. The analysis in **ref.** *3* indicated that the rank test method is asymptotically efficient

From: *Methods in Molecular Biology, vol. 184: Biostatistical Methods*
Edited by: S. W. Looney © Humana Press Inc., Totowa, NJ

when compared to the best parametric tests under a logistic shift model and under most circumstances significantly outperforms tests based on a dichotomized exposure variable, even for fairly small sample sizes. The successful application of the rank test in several case-control studies when the exposure variable was ordinal rather then dichotomized *(1,4,5)* has proved it to be a valuable alternative to more complicated methods, such as multistrata conditional logistic regression. It appears that the rank approach is especially suitable in case-control studies where the exposure variable is poorly characterized but the rank of the exposure of each case among its matching controls is relatively easily established (cf. **Subheading 6.**).

The purpose of this chapter is to examine some properties of the rank-based method within the multistrata case-control study framework. In particular, we are especially concerned here with methods of approximating the power of the appropriate tests for small and moderate sample sizes. This issue is particularly important in the analysis of retrospective/prospective studies of rare diseases (or common diseases of low frequency) when the number of cases is limited. Herein we provide a simple bootstrap algorithm for calculating the approximate power of the appropriate test statistics under translation alternatives. Further, we compare the obtained results with the formulas for exact power of the proposed rank tests under the logistic shift model. The latter is obtained by considering the distributions of appropriate stratum-specific exceedance statistics under the logistic translation alternative. The bootstrap algorithm considered here is somewhat similar to the one for the two-sample Wilcoxon statistic presented by Collings and Hamilton *(6)*, but unlike their algorithm it appears to be consistent.

2. Testing Against a Shift Alternative with Multiple Strata

Suppose that in our retrospective study we have total of n cases of disease under investigation. Here and elsewhere we assume that n is a fixed and, usually, a small number. For instance, in a typical study of rare diseases one would have $n \leq 10$. Corresponding to the ith case $i = 1,...,n$ are n_i controls, usually matched for known or suspected sources of unwanted variation. The case together with its controls form a stratum, and throughout this chapter we assume that there is only one case per stratum and no cases or controls belong to more then one stratum. These assumptions are made mostly for convenience, and the method presented here can be extended to accommodate more complicated study designs (*see*, e.g., Cuzick *[3]*). For each stratum i ($i = 1,...,n$) let R_i be the rank of the exposure of the case among all ($n_i + 1$) individuals in this stratum. There have been at least two tests based on the sum of the R_i's proposed in the literature *(2,3)*. One is based on the statistic

$$W_1 = \sum_{i=1}^{N} R_i \qquad (2.1)$$

and is simply a combination of stratum-specific Wilcoxon two-sample statistics (with respective sample sizes 1 and n_i). The second one is a weighted version of W_1,

$$W_2 = \sum_{i=1}^{N} \frac{R_i}{n_i + 2} \qquad (2.2)$$

By considering the marginal likelihood of ranks for all strata combined, it can be argued that the test based on W_2 is locally most powerful for testing against a logistic shift alternative (cf., e.g., Randles and Wolfe [8, chap. 9]).

Let $F_i(x)$ be the distribution function of the control exposure in the *i*th stratum for $i = 1,...,n$. If we assume that the corresponding case exposure distribution is $F_i(x - \Delta)$, where Δ is the across-strata treatment effect independent of i, then either Eq. (2.1) or Eq. (2.2) could be used to test

$$H_0 : \Delta = 0 \text{ vs } H_1 : \Delta > 0 \qquad (2.3)$$

rejecting H_0 for large values of the statistic. Whereas for small Δ (local alternative) the test based on W_2 would perhaps be preferred over W_1, especially if we have reason to assume that F_i's are logistic cumulative distribution functions (cdf's), the simpler form of W_1 makes it a reasonable competitor for nonlocal alternatives. Obviously, tests based on W_1 and W_2 will coincide when all strata are of equal size.

As in the one-stratum Wilcoxon rank-sum statistic, both W_1 and W_2 are distribution-free under H_0 but depend upon particular forms of the F_i's under H_1. Thus, without some additional assumptions about the forms of the F_i's, there is no closed-form expression for the power of the tests based on **Eq. (2.1)** or **Eq. (2.2)**. Often, even if such assumptions about the stratum exposure distribution are made, the small number of individuals within strata makes their empirical validation (e.g., by means of a goodness-of-fit test) virtually impossible. This is typically the case in the studies of rare diseases with which we are concerned here.

It seems to be of interest, therefore, to introduce a procedure that would allow us to obtain a reasonable approximation of the power of tests based on **Eq. (2.1)** or **Eq. (2.2)** for testing **Eq. (2.3)** without any reference to the particular form of the stratum-specific exposure distributions. This can be accomplished by implementing a bootstrap method similar to that used for approximating the power of the one-stratum Wilcoxon rank-sum test. However, there is one important difference: in our case the bootstrap algorithm will have to perform in a multistrata setting with only one case exposure value per stratum.

3. Bootstrap Algorithm for Estimating Power

Suppose that the amount of the across-strata shift (Δ) between cases and controls is known to be equal either to 0 or some positive constant Δ and let us denote the power of an α-level test ($0 < \alpha < 1$) based on either W_1 or W_2 against a simple alternative $\Delta = d > 0$ by $\Pi(d,\alpha)$. As indicated earlier, $\Pi(\cdot)$ depends also on the F_i's — the exposure distributions for each strata — but for the sake of simplicity we will not reflect this fact in our notation. It should also be noted here that owing to the discrete nature of the test statistics, α may take only finitely many values and thus when performing exact tests we may achieve only finitely many ("natural") α levels.

To approximate $\Pi(d, \alpha)$ we first must obtain estimates of the exposure distributions F_i for $i = 1,...,n$. A number of different approaches are possible here, depending, for instance, on whether we measure exposure on the continuous or ordinal scale. Because in our study of angiosarcoma (cf. **Subheading 6.**) we may take the F_i's to be continuous, we consider the following version of a stratum-specific empirical cdf.

For the i-th stratum let $z_{(1)},... z_{(n_i + 1)}$ be the ordered values of observed exposure levels for all n_i controls belonging to strata i, combined with the translated i-th case exposure level (i.e., the i-th case exposure value minus the quantity d). We assume for convenience that there are no ties among the z's — should that not be the case the procedure described here applies with minor modifications, as long as there are at least two distinct z's. Define $z_{(0)} = 2z_{(1)} - z_{(2)}$ and $z_{(n_i +2)} = 2z_{(n_i + 1)} - z_{(n_i)}$.

Let $\hat{F}_i$ denote the continuous cdf obtained by assigning probability $1/(n_i + 2)$ uniformly over each interval $(z_{(k)}, z_{(k + 1)})$ for $k = 0,...,n_i + 1$. Given the F_i's we may estimate $\Pi(d, \alpha)$ as follows:

1. For each stratum i, draw a computer generated sample of size $n_i + 1$ from $\hat{F}_i$. Add the quantity d to the first observation. This will simulate the shift in the exposure distributions between the case and the controls.

2. Using the first observation as the case value, the statistic W_i ($i = 1, 2$) is calculated from the sample obtained in **step 1**, yielding $W_i^{(0)}$. Let τ be the critical value of the test determined by the condition $\Pi(0, \alpha) = \alpha$. If $W_i^{(0)} \geq \tau$, a success is recorded; otherwise a failure is recorded.

3. Repeat **steps 1** and **2** B times. The bootstrap approximation $\hat{\Pi}(d, \alpha)$ to $\Pi(d, \alpha)$ is given by the binomial proportion of successes among B repetitions. (In all the cases discussed in this chapter, $B \geq 2000$.)

As described in **Subheading 4.**, the simulation study appears to indicate that the above algorithm provides a reasonable estimator of $\Pi(d, \alpha)$ for $n \geq 6$ when the total number of subjects (cases and controls) in all strata combined is at least 36. Asymptotically, the algorithm is consistent, that is, under the sequence of hypotheses $\Delta_N = d_N \rightarrow 0$ such that $\Pi(d_N, \alpha) \rightarrow \text{const} \geq a$, the difference between $\Pi(d_N, \alpha)$ and its bootstrap approximation converges in probability to 0 as the number of cases and controls increases to infinity. This may seem somewhat unusual at first glance, as, in general, the distribution of uncentered U-statistics cannot be approximated consistently by their bootstrapped versions. However, let us note that in the preceding algorithm we do not use a simple bootstrap replica of W_1 (or W_2), but rather its counterpart based on the correctly shifted exposures of the stratum specific case and matching controls.

If the number of cases and controls is not too small (*see* **Subheading 4.**) the bootstrap CLT provides the following alternative approximation of power. After completing **step 1** as in the preceding, in **step 2** we simply calculate the value of W_i, say W_i^0, and then **steps 1** and **2** are repeated B times to obtain the usual approximations to the mean and variance of the bootstrapped version of W_i

$$E^*_B (W_i) = \sum_{b=1}^{B} W_i^0(b)/B$$

$$\mathrm{Var}^*_B (W_i) = \frac{1}{B-1} \sum_{b=1}^{B} \left[W_i^0(b) - E^*_B(W_i) \right]^2 \tag{3.1}$$

The resulting approximate power formula is then given by

$$\Pi(d,\alpha) \approx 1 - \Phi \left(\frac{\tau - E^*_B(W_i)}{\sqrt{\mathrm{Var}^*_B(W_i)}} \right) \tag{3.2}$$

where $\Phi(\cdot)$ stands for the standard normal cdf. The sketch of the mathematical argument supporting this claim can be found in Rempala et al. *(7)*. In this chapter, we are mostly interested in the performance of the bootstrap approximation for a small number of cases (typically about 10) and are less concerned with its large-sample properties. The results of the computer-simulated study of its accuracy for small to moderate sample sizes under the logistic shift model is presented in the next section.

4. Comparison with the Exact Power Under A Logistic Shift Model

Let X_i^0 be the exposure level of the case in stratum i and let $X_i^j, j = 1,...,n_i$ be the levels of the corresponding controls. The X_i^j's for $j = 1,...,n_i$ are therefore distributed according to F_i. To be able to compare the results of the bootstrap power approximation with the true power we have to derive the exact power formula for testing **Eq. (2.3)** using **Eq. (2.1)** or **Eq. (2.2)** and hence impose at this point some additional assumptions on the exposure distributions F_i for $i = 1,...,n$. In what follows we assume the validity of the so called "shift model," that is, we suppose that for each individual exposure level X_i^j we have

$$X_i^j = \Delta\delta_{0i} + \gamma_i + Z_j \tag{4.1}$$

where δ_{0i} is Kronecker's function, γ_i's are some (unknown) positive constants, and Z_j's are independent, identically distributed random variables with any distribution. For the purpose of the computer simulation described in this subheading we have taken the Z_j's to be logistic $L(0,1)$ random variables. With this particular choice of the Z_j's, we refer to **Eq. (4.1)** throughout the paper as "the logistic shift model" (cf. also Cuzick *[3]*).

The calculation of the exact power of the location shift test (**Eq. [2.3]**) based on **(2.1)** or **(2.2)** under **Eq. (4.1)** with the Z_j's being logistic random variables can be accomplished by considering the distributions of stratum-specific exceedance statistics (cf. Katzenbeisser *[9,10]*) and then combining the results by multiple convolution. Plots of the distributions of the statistic **Eq. (2.1)** and **Eq. (2.2)** under the logistic shift model with eight cases and number of controls per stratum as in the angiosarcoma study (*see* **Subheading 6.**) for the shift values $\Delta = 0, 1, 2, 3$ are presented in **Figs. 1** and **2**. The general formula for the distribution of **Eqs. (2.1)** and **(2.2)** in terms of exceedance statistics under the logistic shift model is provided in Rempala et al. *(7)*. As can be seen from the plots the normal approximation is fairly accurate even for large values of Δ. In fact, the normal approximation works reasonably well for Δ between 0 and 3 for statistics W_1 and W_2 (for W_1 we need to apply a continuity correction) as long as the number of strata is at least 6, the number of controls per stratum is at least 3, and the total number of controls is at least 36.

Having obtained the exact distribution of statistics (**Eqs. [2.1]** and **[2.2]**), for logistic shifts, we may calculate the true power of tests based on W_1 or W_2, which in turn allows us to assess the accuracy of our bootstrap approximation algorithm. Such a comparison with eight cases and numbers of controls coinciding with the numbers from the angiosarcoma study described in **Subheading 6.** is given in **Fig. 3**. As can be seen from the plot, for the logistic shift model, the overall performance of our bootstrap approximation appears to be quite satisfactory even for a relatively small number of cases ($n = 8$).

5. Estimation of the Across-Strata Treatment Effect

So far we have assumed that the amount of the across-strata shift between the exposure distributions of cases and controls (Δ) is known. In practical situations this is obviously rarely the case and we need a way to estimate Δ. One of the standard nonparametric approaches would be to take

$$\hat{\Delta} = \frac{1}{n} \sum_{i=1}^{n} \left(X_i^0 - \bar{X}_i \right) \tag{5.1}$$

but we do not propose to do so because the above estimator of Δ is nonrobust against outliers and for small sample sizes may be quite misleading. More robust estimators, say $\hat{\Delta}_1$ and $\hat{\Delta}_2$, may be obtained by using the Hodges–

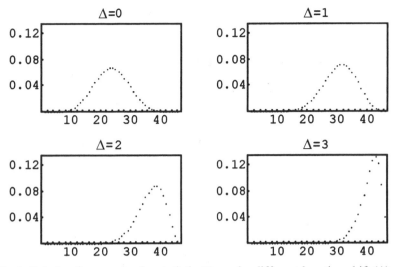

Fig. 1. Relative frequencies for statistic W_1, under different location shift (Δ) values, with eight cases and the number of controls per case equal to 2, 3, 3, 4, 7, 8, 8, 10, respectively. The Central Limit Effect is clearly visible even for moderate $\Delta \leq 2$.

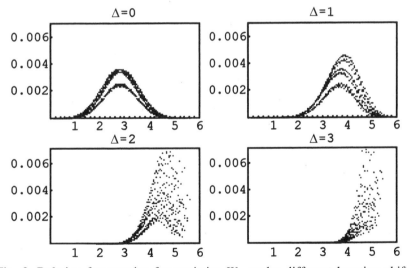

Fig. 2. Relative frequencies for statistics W_2, under different location shift (Δ) values, with eight cases and the number of controls per case equal to 2, 3, 3, 4, 7, 8, 8, 10, respectively. The Central Limit Effect is clearly visible even for moderate $\Delta \leq 2$.

Lehmann method *(11)* based on the statistics **Eqs. (2.1)** and **(2.2)**, respectively. The use of the Hodges–Lehmann technique is appropriate here because, under the null hypothesis, W_1 and W_2 are symmetric about

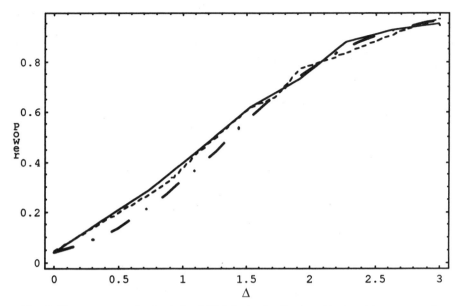

Fig. 3. True power under logistic shift *(chain-dot line)* and its bootstrap approximations using binomial proportion *(solid line)* and bootstrap CLT *(dashed line)* for test
statistic W_1 with eight cases and number of controls per case equal to 2, 3, 3, 4, 7, 8, 8,
10, respectively.

$n + \sum_{i=1}^{n} n_i / 2$ and $n + \sum_{i=1}^{n} n_i / (2n_i + 4)$, respectively. To find the explicit forms
for the estimators, let us define $D_{ij} = X_i^0 - X_i^j$ for $i = 1,...,n$ and $j = 1,...,n_i$ to be
the within-stratum exposure level differences between the case and matched
controls. Then, using the Hodges–Lehmann argument (cf., e.g., Randles and
Wolfe [*8*, p. 208]), it is easy to show that

$$\hat{\Delta}_1 = med(D_{ij}) \qquad (5.2)$$

the median of the D_{ij}'s. Similar reasoning applies to $\hat{\Delta}_2$ although now we have
to consider a "weighted version" of the D_{ij}'s, due to the weights present in
Eq. (2.2). Namely, let n* = $\Pi_{i=1}^{n} (n_i + 2)$ and let us consider the extended list of
D_{ij}'s in which each D_{ij} is repeated exactly n*/$(n_i + 2)$ times. Then

$$\hat{\Delta}_1 = med \left(\text{extended list of } D_{ij}\right) \qquad (5.3)$$

Obviously, **Eqs. (5.2)** and **(5.3)** coincide when all n_i are equal. If the
exposure distributions F_i's are symmetric, then $\hat{\Delta}_1$ and $\hat{\Delta}_2$ are unbiased for Δ.
Otherwise, under most circumstances they are median unbiased (or almost so).
In our setting, **Eqs. (5.2)** and **(5.3)** appear to be more appropriate estimators of
the treatment effect than **Eq. (5.1)**. The estimators of the standard errors of

Eqs. (5.2) and **(5.3)** may again be obtained using the bootstrap method in a fashion similar to **Eq. (3.1)**. Under most circumstances, the standard errors for $\hat{\Delta}, \hat{\Delta}_1, \hat{\Delta}_2$ and will be of order $O(n^{-1})$ and hence, for small n, the variability of all three estimators could be quite high (cf. **Subheading 6.**). In such cases, in addition to directly estimating Δ one might also wish to estimate $\theta_i = P(X_i^0 > X_i^j)$, the probability that the i-th case exposure exceeds that of its control. This can be easily done, since the quantity $(R_i - 1)/n_i$ is always an unbiased estimator of θ_i. Under the shift model (**Eq. [4.1]**) this parameter, say θ, is the same in all strata and may be estimated unbiasedly by the statistics (cf. also Cuzick *[3]*):

$$\hat{\theta}_1 = \frac{\sum_{i=1}^{N} (R_i - 1)}{\sum_{i=1}^{N} n_i} = \frac{W_1 - n}{\sum_{i=1}^{N} n_i}$$

and

$$\hat{\theta}_2 = \frac{\sum_{i=1}^{N} (R_i - 1)/(n_i + 2)}{\sum_{i=1}^{N} n_i/(n_i + 2)} = \frac{W_2 - \sum_{i=1}^{N} 1/(n_i + 2)}{\sum_{i=1}^{N} n_i/(n_i + 2)}$$

Because $\hat{\theta}_1$ and $\hat{\theta}_2$ are both linear combinations of stratum specific Wilcoxon–statistics their standard errors will typically be much smaller than that of **Eqs. (5.1)–(5.3)**. Under **Eq. (4.1)** the parameter θ can sometimes be expressed as an explicit function of Δ. For instance, if Δ is close to zero the delta method shows that

$$\Delta \approx \frac{\theta - 1/2}{\int f^2 (x)\, dx}$$

provided that $\int f^2 (x)\, dx < \infty$, where $f(x)$ is the density function of F_1. Under the logistic shift model, for instance, this yields $\Delta \approx 6\theta - 3$.

6. A Study of Angiosarcoma Occurrences Among Chemical Industry Workers

In this subheading, we illustrate the method with an example that has in fact motivated our study of multistrata rank-based methods. The data presented were collected over the past 20 yr by researchers from the Division of Occupational Toxicology at the University of Louisville School of Medicine as part of an effort to examine the relationship between occupational exposure to suspected carcinogens and the development of disease. In 1974, in response to the discovery of cases of hepatic angiosarcoma among its employees, the B. F. Goodrich Louisville Chemical Plant jointly with the University of Louisville School of Medicine, developed an exposure monitoring system utilizing rank ordering of exposures for highly suspected chemicals. The exposure index combined two components: work history and a job exposure category based on a 7-point scale that rated the monthly exposure of each employee on any particular job from 0 (absent from the environment) to 6 (very frequent intimate skin contact or high inhalation). The exposure levels for different jobs performed during the month by a given employee were then weighted by their duration to give the total monthly exposure index for that employee. This index was accumulated across months of employment to give the cumulative exposure rank months (CERM). Obviously, CERM is, by its design, a very imprecise measure of exposure but was thought to be the best available, because only a minimal amount of information concerning historical exposures could be found in the company records. Thus, owing to the nature of CERM the standard analysis based on the conditional likelihood method and logistic regression or the hypergeometic distribution (as offered, for instance, by StatXact) could yield misleading results. The rank method based only on the relative magnitude of exposures seems to be more appropriate.

In **Tables 1** and **2**, we present the total stratum-standardized CERM for the exposure to the chemical vinyl chloride, along with corresponding rank-statistics, for the eight cases of angiosarcoma identified among B. F. Goodrich Chemical Plant employees between the time of the first case (diagnosed in January, 1974) and January, 1998, together with the total stratum-standardized CERM for the controls matched by sex, age, and length of employment. Standardization of CERM within any given strata is obtained by dividing the raw CERM values by the stratum-specific CERM standard deviation for the controls. Of course, the underlying assumption here is that, within-strata, exposures of the case and its corresponding controls are measured on the same scale.

It is easily seen from the values of the statistics reported in **Table 2** that the data show significant association of the exposure to vinyl chloride (as measured in CERM) with the development of hepatic angiosarcoma. When testing **Eq. (2.3)** by means of **Eq. (2.1)** and **Eq. (2.2)**, we find the normal approximations of the P-values to be 0.01262 and 0.01252, respectively. The achieved power of the tests is estimated to be about 0.60 for W_1 and about 0.65 for W_2 at the 5% significance level.

7. Summary

The rank method of analyzing multistrata case-control studies was considered. The consistent bootstrap algorithm for approximating the power of the rank-based test was presented. The computer simulation study presented in this chapter indicates that, under a logistic shift alternative, the bootstrap algorithm provides a reasonably good approximation of the true power, even for relatively small sample sizes, provided that the across-strata shift (treatment effect) is known. For an unknown shift, a consistent and robust estimator of the Hodges–Lehmann type was proposed.

Table 1
Values of the Total Stratum-Standardized CERM for Vinyl Chloride of the Eight Angiosarcoma Cases and Their Matched Controls Not Developing the Disease Among the Employees of B. F. Goodrich Chemical Plant (Matching Done by Sex, Age, and the Length of Employment)

Case value	Number of controls (n_i)	Control values	Case rank (R_i)
3.91	8	2.14, 2.66, 2.76, 2.99, 4.39, 1.13, 1.89, 4.41	7
4.05	7	2.41, 2.39, 4.04, 0.98, 1.69, 3.27, 2.07	8
2.76	8	2.99, 1.82, 1.8, 3.63, 2.00, 1.18, 2.33, 2.85	6
4.21	10	2.4, 1.56, 3.27, 1.93, 2.3, 1.09, 2.53, 4.56, 3.59, 2.96	10
5.78	3	5.25, 4.5, 3.36	4
1.87	2	1.18, 0.82	3
2.82	4	1.24, 3.57, 1.63, 2.63	4
4.85	3	5.16, 3.13, 5.23	2

Table 2
The Values of Test Statistics W_1 and W_2, Along with the Corresponding Estimates of the Shift (Δ), Mean Percentile Shift (θ), and Achieved Power

	W_i	P-Value	$\hat{\Delta}$ (s.e.)	$\hat{\Delta}_i$ (s.e.)	$\hat{\theta}_i$ (s.e.)	$\hat{\Pi}(\hat{\Delta}_i, 0.05)$
$i = 1$	43.00	0.01262	1.03 (0.20)	1.13 (0.15)	0.8 (0.056)	0.60
$i = 2$	5.53	0.01252	1.03 (0.20)	1.18 (0.10)	0.79 (0.05)	0.65

References

1. Greenberg, R. A. and Tamburro, C. H. (1981) Exposure indices for epidemiological surveillance of carcinogenic agents in an industrial chemical environment. *J. Occup. Med.* **23**, 353–358.
2. van Elteren, P. (1960) On the combination of independent two-sample tests of Wilcoxon. *Bull. Int. Statist. Instit.* **37**, 351–361.
3. Cuzick, J. (1985) A method for analyzing case-control studies with ordinal exposure variables. *Biometrics* **41**, 609–621.

4. Cuzick, J., Bulstrode, J. C., Stratton, I.,Thomas, B. S., Bulbrook, R. D., and Hayward, J. L. (1983) A prospective study of urinary androgen levels and ovarian cancer. *Int. J. Cancer* **32,** 13–19.

5. Kwa, H. G., Cleton, F., Wang, D. Y., Bulbrook, R. D., Bulstrode, J. C., Hayward, J. L., et al. (1981) A prospective study of plasma prolactin levels and subsequent risk of breast cancer. *Int. J. Cancer* **28,** 673–676.

6. Collings, B. J. and Hamilton, M. A. (1988) Estimating the power of the two-sample Wilcoxon test for location shift. *Biometrics* **44,** 847–860.

7. Rempala, G., Looney, S., Tamburro, C., and Fortwengler, P (1998) *Power of the rank test for multi-strata case-control studies.* University of Louisville Department of Mathematics Technical Report no. 1/98.

8. Randles, R. H. and Wolfe, D. A. (1979) *Introduction to the Theory of Nonparametric Statistics.* Krieger, Malabar, FL.

9. Katzenbeisser, W. (1985) The distribution of two-sample location exceedance statistic under Lehmann alternatives. *Statistische Hefte* **26,** 131–138.

10. Katzenbeisser, W. (1989) The exact power of two-sample location tests based on exceedance statistics against shift alternatives. *Statistics* **20,** 47–54.

11. Hodges, J. L. and Lehmann, E. L. (1963) Estimates of location based on rank tests. *Ann. Math. Statist.* **34,** 598–611.

Index

About the Editor

Stephen W. Looney has over 20 years experience as an applied statistician. He is currently Professor of Family and Community Medicine and a Biostatistician at the University of Louisville School of Medicine in Louisville, Kentucky where he has served on the faculty since 1991. Prior to that, he was a member of the faculty in the Department of Quantitative Business Analysis at Louisiana State University in Baton Rouge (1981–1991), and worked as Chief Statistical Analyst at the Northeast Health District in Athens, Georgia (1979–1980). He has held visiting positions at Keele University in England, and at The Upjohn Company, Argonne National Laboratory, Oak Ridge National Laboratory, Health and Welfare Canada, and the National Center for Health Statistics.

Dr. Looney became interested in biomarkers through his work in the Center for Environmental Health Sciences at the University of Louisville. This research focused on the detection of residential and environmental exposures to hazardous chemicals, primarily acrylonitrile. Dr. Looney's other research interests include health outcomes research, statistical inference, and statistical computing. He is the author or co-author of 40 articles in scientific journals and has participated in over 100 presentations at professional meetings.

Born and raised in northeast Georgia, Dr. Looney received his Ph.D. in Statistics from The University of Georgia in 1980, and his B.S. in Mathematics from the same institution in 1974, at which time he graduated summa cum laude with honors. He also received an M.S. in Mathematics from the University of Virginia in 1976. Dr. Looney is a member of Phi Beta Kappa, Phi Kappa Phi, and Phi Eta Sigma honorary societies.

Dr. Looney was elected a Fellow of the American Statistical Association in 1998 and a Fellow of the Royal Statistical Society in 1997. He has also been selected for membership in Sigma Xi Research Society, Who's Who in Science and Engineering, and Who's Who in Technology. Dr. Looney has received awards for his teaching from the Louisiana State University College of Business Administration and the L.S.U. Student Government Association.